Synthesis Lectures on Data Management

Series Editor

H. V. Jagadish, University of Michigan, Ann Arbor, MI, USA

This series publishes lectures on data management. Topics include query languages, database system architectures, transaction management, data warehousing, XML and databases, data stream systems, wide scale data distribution, multimedia data management, data mining, and related subjects.

Shaoxu Song · Lei Chen

Integrity Constraints on Rich Data Types

 Springer

Shaoxu Song
School of Software
Tsinghua University
Beijing, China

Lei Chen
Department of Computer Science
and Engineering
Hong Kong University of Science
and Technology
Hong Kong, Hong Kong

ISSN 2153-5418   ISSN 2153-5426 (electronic)
Synthesis Lectures on Data Management
ISBN 978-3-031-27179-3   ISBN 978-3-031-27177-9 (eBook)
https://doi.org/10.1007/978-3-031-27177-9

This Springer imprint is published by the registered company Springer Nature Switzerland AG
The registered company address is: Gewerbestrasse 11, 6330 Cham, Switzerland

Acknowledgements

We would like to thank Dr. Fei Gao, Dr. Ruihong Huang and Dr. Chaokun Wang for the discussions in the early stage of this study. This work is supported in part by the National Natural Science Foundation of China (62072265, 62232005, 62021002), the National Key Research and Development Plan (2021YFB3300500, 2019YFB1705301, 2019YFB1707001), Beijing National Research Center for Information Science and Technology (BNR2022RC01011), and Alibaba Group through Alibaba Innovative Research (AIR) Program.

Contents

1 Introduction 1
   1.1 Background 3
   1.2 Motivation 4
   1.3 Categorization on Data Types 4
   1.4 Perspectives on Data Dependencies 5
   1.5 Related Studies in Dependency Survey 11
   1.6 Organization 12

2 Categorical Data 15
   2.1 Functional Dependencies (FDs) 16
   2.2 Equality Generating Dependencies (EGDs) 19
   2.3 Soft Functional Dependencies (SFDs) 21
   2.4 Probabilistic Functional Dependencies (PFDs) 23
   2.5 Approximate Functional Dependencies (AFDs) 25
   2.6 Numerical Dependencies (NUDs) 26
   2.7 Conditional Functional Dependencies (CFDs) 27
   2.8 Extended Conditional Functional Dependencies (eCFDs) 30
   2.9 Multivalued Dependencies (MVDs) 32
   2.10 Full Hierarchical Dependencies (FHDs) 34
   2.11 Approximate Multivalued Dependencies (AMVDs) 37
   2.12 Inclusion Dependencies (INDs) 39
   2.13 Approximate Inclusion Dependencies (AINDs) 41
   2.14 Conditional Inclusion Dependencies (CINDs) 43
   2.15 Summary and Discussion 44

3 Heterogeneous Data 47
   3.1 Metric Functional Dependencies (MFDs) 48
   3.2 Neighborhood Dependencies (NEDs) 50
   3.3 Differential Dependencies (DDs) 52
   3.4 Conditional Differential Dependencies (CDDs) 54
   3.5 Comparable Dependencies (CDs) 56

3.6   Probabilistic Approximate Constraints (PACs) 58
3.7   Fuzzy Functional Dependencies (FFDs) 60
3.8   Ontology Functional Dependencies (ONFDs) 63
3.9   Matching Dependencies (MDs) 65
3.10 Conditional Matching Dependencies (CMDs) 67
3.11 Summary and Discussion 68

4 Ordered Data 71
4.1   Ordered Functional Dependencies (OFDs) 71
4.2   Order Dependencies (ODs) 74
4.3   Band Order Dependencies (BODs) 76
4.4   Denial Constraints (DCs) 78
4.5   Sequential Dependencies (SDs) 80
4.6   Conditional Sequential Dependencies (CSDs) 81
4.7   Summary and Discussion 83

5 Temporal Data 85
5.1   Temporal Functional Dependencies (TFDs) 86
5.2   Trend Dependencies (TDs) 88
5.3   Speed Constraints (SCs) 90
5.4   Multi-speed Constraints (MSCs) 92
5.5   Acceleration Constraints (ACs) 95
5.6   Temporal Constraints (TCs) 97
5.7   Petri Nets (PNs) 99
5.8   Summary and Discussion 101

6 Graph Data 103
6.1   Neighborhood Constraints (NCs) 104
6.2   Node Label Constraints (NLCs) 106
6.3   Path Label Constraints (PLCs) 108
6.4   XML Functional Dependencies (XFDs) 109
6.5   XML Conditional Functional Dependencies (XCFDs) 111
6.6   Keys for Graph (GKs) 113
6.7   Graph-Patterns Association Rules (GPARs) 115
6.8   Functional Dependencies for Graph (GFDs) 116
6.9   Graph Entity Dependencies (GEDs) 118
6.10 Graph Association Rules (GARs) 121
6.11 Graph Differential Dependencies (GDDs) 123
6.12 Graph Denial Constraints (GDCs) 125
6.13 Temporal Dependencies for Graph (TGFDs) 126
6.14 Summary and Discussion 128

7 Conclusions and Directions 131

Index of Data Dependencies 135

References 137

About the Authors

Shaoxu Song is an associate professor in the School of Software, Tsinghua University, Beijing, China. His research interests include data quality and data integration. He has published more than 50 papers in top conferences and journals such as SIGMOD, VLDB, ICDE, TODS, VLDBJ, TKDE, etc. He served as a vice program chair for the 2022 IEEE International Conference on Big Data (IEEE BigData 2022). He received a Distinguished Reviewer award from VLDB 2019 and an Outstanding Reviewer award from CIKM 2017.

Lei Chen is a Chair Professor in the Department of Computer Science and Engineering, Hong Kong University of Science and Technology, and the Director of HKUST Big Data Institute. He received a SIGMOD Test-of-Time Award in 2015 and served as PC Co-Chair of VLDB 2019 and ICDE 2023. He is currently the Editor-in-Chief of the VLDB Journal and the Editor-in-Chief of IEEE Transactions on Knowledge and Data Engineering (TKDE). He is an IEEE Fellow and ACM Distinguished Scientist.

Introduction

Data dependencies, such as *functional dependencies* (FDs), have been long recognized as integrity constraints in databases [42]. They are firstly utilized in database design [3]. Conventionally, data dependencies are extracted from application requirements for database standardization, and used in database design to ensure data quality. For instance, functional dependencies (FDs) [16] are employed to evaluate whether a relation is in third normal form (3NF) [42] or Boyce-Codd normal form (BCNF) [43]. FDs are extended to *multivalued dependencies* (MVDs) [13], i.e., every FD is also an MVD, in order to test whether a relation is in fourth normal form (4NF) [55]. Such data dependencies are also used in database query optimization [53, 86].

While data dependencies and their extensions are conventionally used for schema design, they have been recently revisited and extended for big data analysis. When the users are exploring data, they often experience the variety and veracity issues. The variety and veracity issues in big data bring challenges for applications of data dependencies. As mentioned in [51, 139], the variety issue means that constraints may apply only to certain subsets of data, and the veracity issue refers to data quality problems such as inconsistency and invalidity. To process the possibly dirty data from heterogeneous sources, data dependencies have been recently extended and used for improving data quality [63] such as error detection [163], data repairing [23, 110], data deduplication [70], etc.

Given the various data types, ranging from the conventional categorical data, numerical data to the more prevalent heterogeneous data, different extensions are made over the data dependencies with distinct expressive power. In this study, we propose to give an entire landscape of typical data dependencies, in order to identify their relationships and distinct application scenarios. For example, if a user wants to perform data repairing over a data source with both categorical and numerical values, a direct suggestion will be *denial constraints* (DCs) [159] referring to Fig. 1.1.

S. Song and L. Chen, *Integrity Constraints on Rich Data Types*,
Synthesis Lectures on Data Management,
https://doi.org/10.1007/978-3-031-27177-9_1

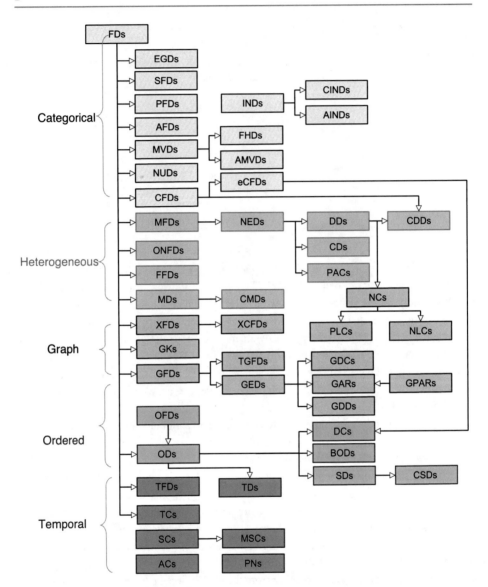

Fig. 1.1 A family tree of extensions between various data dependencies

1.1   Background

An FD $X \rightarrow Y$, over relation R, where $X, Y \subseteq R$, states that if any two tuples in an instance of R have equal X-values, then their Y-values should also be identical. Such integrity constraints can be used to improve the quality of the database schema and reduce operational abnormalities. In practice, data is often dirty, that is, it contains inconsistencies, conflicts and errors with more costs and risks. In addition, the widespread use of the Web makes it possible to extract and integrate data from various sources on an unprecedented scale, but it also increases the risk of creating and disseminating dirty data. When we are extracting millions of relational HTML tables from the Web, traditional data integration methods cannot be extended to a corpus with hundreds of small tables in a domain. To solve this problem, we can use a framework based on functional dependencies. By analyzing dependencies in data, errors can be identified and inconsistencies can be located. Data dependencies should play an important role in data quality activities and tools. They can specify the semantics of dependent data and capture inconsistencies and errors due to violations of data dependencies.

Consider the example relation instance $r_{1.1}$ illustrated in Table 1.1. A functional dependency fd_1 below over $r_{1.1}$ specifies the constraint that for any two tuples of hotels, if they have the same address, then their region values must be equal,

$$\mathsf{fd}_1 : \mathrm{address} \rightarrow \mathrm{region}.$$

For instance, tuples t_1 and t_2 in Table 1.1, with the same address value "No.5, Central Park", have equal region value "New York" too. That is, each address is associated with precisely one region.

Table 1.1 An example relation instance $r_{1.1}$ of hotel

	Name	Address	Region	Star	Price
t_1	New Center	No.5, Central Park	New York	3	299
t_2	New Center Hotel	No.5, Central Park	New York	3	299
t_3	St. Regis Hotel	#3, West Lake Rd.	Boston	3	319
t_4	St. Regis	#3, West Lake Rd.	Chicago, MA	3	319
t_5	West Wood Hotel	Fifth Avenue, 61st Street	Chicago	4	499
t_6	West Wood	Fifth Avenue, 61st Street	Chicago, IL	4	499
t_7	Christina Hotel	No.7, West Lake Rd.	Boston, MA	5	599
t_8	Christina	#7, West Lake Rd.	San Francisco	5	0

This fd_1 can be used to detect data quality issues in the relation instance $r_{1.1}$ in Table 1.1. For tuples t_3 and t_4 with equal values on address, they have different region values, which are then treated as a violation to the above fd_1. It implies errors occurred in t_3 or t_4, e.g., "Chicago, MA" should be "Boston" instead.

1.2   Motivation

Owing to the variety issue of big data, real-world information often has various representation formats. As indicated in [157], the strict equality restriction limits the usage of FDs. Therefore, data dependencies notations often need adaption to meet the requirements of various data types. Since the research on integrity constraints starts in the early 1970s, some methods in the previous literature research have been developed in many versions. This requires further review of these methods. Because various data dependencies have different characteristics, advantages and applications, this study can help researchers in the fields of databases, data quality, data cleaning, machine learning and knowledge discovery to select more suitable methods.

For example, according to fd_1, tuples t_5 and t_6 in Table 1.1 will be detected as a "violation", since they have "different" region values but the same address. However, "Chicago" and "Chicago, IL" indeed denote the same region in the real-world with different representation formats, i.e., no errors. On the other hand, t_7 and t_8, which have similar addresses values but different regions, are true violations with data errors. Unfortunately, they cannot be detected by fd_1, since their address values are not exactly equal (but similar). The fd_1 considers only those tuples with the strict equality relationships on address.

Therefore, data dependencies notations often need adaption to meet the requirements of various data types.

1.3   Categorization on Data Types

In this book, motivated by the aforesaid veracity and variety issues of big data, we focus on recent proposals of novel data dependencies declared over various data types. Each chapter will discuss one of the following types. Appendix A presents a full list of acronyms for data dependencies.

(1) For conventional **categorical data**, it is noticed that data dependencies might no longer hold over the entire set of all tuples. For example, as illustrated in Table 1.1, while t_1 and t_2 satisfy fd_1, t_5 and t_6 (which contain no error) do not. To support such scenarios, the important attempts are to extend data dependencies with conditions [22] or statistics [98]. The basic idea of these extensions is to make the dependencies, that originally hold for the whole table, valid only for a subset of tuples.

(2) The aforesaid extensions over categorical data still consider the equality relationship of data values. This strict constraint on equality limits the usage of data dependencies over **heterogeneous data**, since real-world information often has various representation formats or conventions, such as "Chicago" and "Chicago, IL" in Table 1.1. To improve the expressive power, distance metrics are introduced to data dependencies [146]. Instead of equality, data dependencies with distance/similarity metrics can specify constraints on "(dis)similar" semantics. For example, a distance constraint could state that if two tuples have similar address values, then they are similar on regions as well.

(3) In addition to categorical and heterogenous data, another typical data type is **ordered data**, such as star and price in Table 1.1. Order relationships are usually more important than the equality relationships for numerical data [50]. For instance, as illustrated in Table 1.1, a higher end hotel generally has a higher price.

(4) As a special type of numerical data, **temporal data** are often associated with timestamps [1]. They are usually generated continuously by machines or sensors and collected as time series, e.g., hourly temperature, or event traces such as flight arrival and departure events. Again, integrity constraints could be declared over such temporal data, for instance, specifying that the temperature changes in a day should be bounded [158].

(5) With the development of social networks and knowledge bases, **graph data** have also been extensively studied. While some structure information is explicitly presented, in the form of extensive markup language (XML) or resource description framework (RDF), other graph structures could be implicitly implied, e.g., referring to the similarity relationships between tuples. The integrity constraints are specified over the vertexes, edges and labels in the graph, such as the neighborhood constraints [153].

1.4 Perspectives on Data Dependencies

For each data dependency, in addition to the definition and example of the dependencies notation, we are particularly interested in four perspectives, including (a) extension relationships between data dependencies, how they generalize other data dependencies; (b) axiomatization of dependencies, how they imply other data dependencies; (c) discovery of data dependencies from data instances, how they can be obtained from data; and (d) application of data dependencies in data quality tasks, how they are utilized.

In particular, the extension relationships among various data dependencies, mostly rooted in functional dependencies (FDs), form a family tree as presented in Fig. 1.1. In addition, we provide the time when the data dependency was first proposed in Fig. 1.2. The comparison of the importance of various data dependencies according to the number of publications using them is also shown in Fig. 1.3.

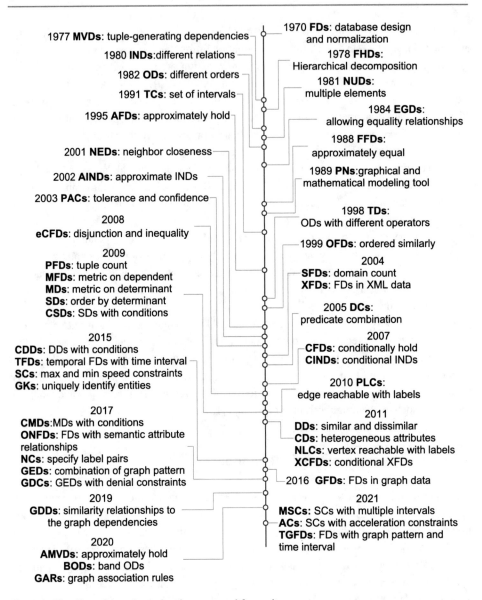

1977 **MVDs**: tuple-generating dependencies

1970 **FDs**: database design and normalization

1980 **INDs**:different relations

1978 **FHDs**: Hierarchical decomposition

1982 **ODs**: different orders

1981 **NUDs**: multiple elements

1991 **TCs**: set of intervals

1984 **EGDs**: allowing equality relationships

1995 **AFDs**: approximately hold

1988 **FFDs**: approximately equal

2001 **NEDs**: neighbor closeness

1989 **PNs**:graphical and mathematical modeling tool

2002 **AINDs**: approximate INDs

2003 **PACs**: tolerance and confidence

1998 **TDs**: ODs with different operators

2008
eCFDs: disjunction and inequality

1999 **OFDs**: ordered similarly

2009
PFDs: tuple count
MFDs: metric on dependent
MDs: metric on determinant
SDs: order by determinant
CSDs: SDs with conditions

2004
SFDs: domain count
XFDs: FDs in XML data

2005 **DCs**: predicate combination

2015
CDDs: DDs with conditions
TFDs: temporal FDs with time interval
SCs: max and min speed constraints
GKs: uniquely identify entities

2007
CFDs: conditionally hold
CINDs: conditional INDs

2010 **PLCs**: edge reachable with labels

2017
CMDs:MDs with conditions
ONFDs: FDs with semantic attribute relationships
NCs: specify label pairs
GEDs: combination of graph pattern
GDCs: GEDs with denial constraints

2011
DDs: similar and dissimilar
CDs: heterogeneous attributes
NLCs: vertex reachable with labels
XCFDs: conditional XFDs

2016 **GFDs**: FDs in graph data

2019
GDDs: similarity relationships to the graph dependencies

2021
MSCs: SCs with multiple intervals
ACs: SCs with acceleration constraints
TGFDs: FDs with graph pattern and time interval

2020
AMVDs: approximately hold
BODs: band ODs
GARs: graph association rules

Fig. 1.2 Timeline of data dependencies proposed for various reasons

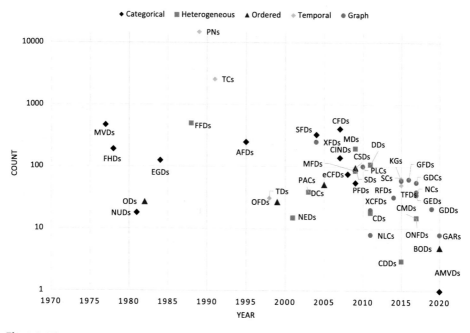

Fig. 1.3 The number of publications mentioning data dependencies

Family Tree of Extensions

To capture the semantics of big data with variety issues, data dependencies are extended with more expressive power. In order to choose data dependencies with proper expressive power in practice, the complicated extension relationships between data dependencies need to be investigated. Figure 1.1 presents the extension relationships, where each arrow denotes the extension/generalization relationship between two types of data dependencies. For example, an arrow from FDs to SFDs denotes that SFDs extend/generalize/subsume FDs. That is, SFDs are the extension of FDs, SFDs are the generalization of FDs, SFDs subsume the semantics of FDs, or FDs are special SFDs (with strength 1).[1] Each extension relationship is explained in the following sections as a special case of data dependencies. Appendix A presents an index of all the studied data dependencies. Again, Fig. 1.1 categorizes all the data dependencies into five main parts, denoted by different colors, following the categorization in Sect. 1.3. It is notable that data dependencies in five parts may interact. For example, most dependencies are generalizations/extensions of FDs, which serve as the root of the whole family tree. Moreover, as presented in Sect. 3.4, CDDs, an extension of CFDs, apply to a relation with both categorical and heterogeneous data. Similarly, DCs are extensions of eCFDs, applying to both categorical and numerical data. A special case of numerical data is a timestamp, where temporal constraints on timestamps are studied, such as sequential dependencies (SDs) [87]

---

[1] See Sect. 2.3 for details.

and speed constraints (SCs) [158]. The neighborhood constraints (NCs) extend DDs, when the graphs are built on similarity networks. To sum up, data dependencies could be applied to a dataset with rich data types, i.e., categorical, heterogeneous, numerical, temporal and graph data.

Figure 1.2 provides a timeline of data dependencies that are proposed for various reasons. There are some important milestones. AFDs [109] are first proposed in 1995 for data dependencies that approximately hold in a relation. Similar extensions such as SFDs [98] in 2004 and PFDs [167] in 2009 are developed following the same line. Moreover, CFDs introduce another series of extensions on conditionally holding in a relation, e.g., CSDs [22] in 2009, CDDs [116] in 2015 and CMDs [175] in 2017.

In order to compare the impact of data dependencies, Fig. 1.3 illustrates the number of publications mentioning data dependencies, according to Google Scholar. As shown, while the extensions over the conventional categorical data such as CFDs attract more attention, recent proposals focus more on the heterogeneous data, e.g., MDs, DDs and their extensions. Moreover, the usage of data dependencies over ordered data is increasing, from canonical ODs to recent SDs.

Axiomatization of Dependencies

Axioms are a set of inference rules that reveal insights of logical implication and can be used for dependency discovery and symbolic proofs. Armstrong's axioms for FDs, developed by Armstrong [8], are the most well-known axioms, consisting of three rules: Reflexivity, Augmentation and Transitivity. Armstrong's axioms guide the establishment of axioms for other dependencies later on.

For the class of categorical data, analogous to Armstrong's Axioms, an inference system is provided for CFDs [65], where two inference rules extend Armstrong's axioms of Reflexivity and Augmentation and the others are specific to CFDs. Four rules constitute a complete set of inference rules for MVDs [13], including Complementation, Reflexivity, Augmentation and Transitivity.

For the class of heterogeneous data, Song and Chen [146] introduce a complete and sound set of inference rules for DDs. There are four inference rules defined for DDs, three of which are modified according to the Reflexivity, Augmentation and Transitivity rules in Armstrong's axioms, and one is different from Armstrong's axioms and specific to differential functions.

For the class of ordered data, following Armstrong's axiom system, Ng [134] gives a set of inference rules for POFDs. The axiom system comprising from four rules (Reflexivity, Augmentation, Transitivity and Permutation) is complete and sound.

For the class of temporal data, Wijsen [176] provides a complete and sound axiomatization of logical implication for TDs, where two rules (Augmentation and Transitivity) extend Armstrong's axioms.

For the class of graph data, GEDs are finitely axiomatizable and the set of inference rules $\mathcal{A}_{GED}$ is sound, complete and independent (non-redundant and minimal) [73]. $\mathcal{A}_{GED}$ contains six inference rules and can derive Reflexivity, Augmentation and Transitivity, which means Armstrong's axioms also hold for GEDs.

Discovery from Data Instance

Data dependencies are often specified directly by the database designer or extracted from business requirements. That is, the semantics of data dependencies are studied before the data ingestion into databases. In contrast, in the big data era, rich data types are collected from existing sources with weakly specified integrity constraints. While manually specifying constraints is no longer an option over such big data, it often relies on automatically discovering data dependencies from the data.

The discovery problem is to find the data dependencies that hold, or are expected to hold, in a given dataset (either clean or dirty). For instance, fd_1 is found as an FD that (approximately) holds in Table 1.1. The discovery of FDs from data is known to be intrinsically hard, i.e., a minimal cover can be exponentially large with respect to the number of attributes in a relation [128, 129, 143]. The problem of determining whether a relation has a key of size less than a given integer is NP-complete [12]. Efficient strategies have been designed for discovering FDs, such as TANE [95, 96], FastFD [180]. FD discovery targets on generating a minimal cover of all FDs that hold in the given data.

Fan et al. [66] study the discovery of CFDs and propose three algorithms for different scenarios, including an extension to the level-wise TANE algorithm. Chiang and Miller [36] also study the discovery of CFDs in a level-wise style. It is thereby not surprising that discovering data dependencies extending FDs is generally costly.

While the extensions with metrics enable data dependencies with more expressive power and tolerance to variations of heterogeneous data, the data dependencies become more complicated. In particular, the thresholds of metric distances/similarities are often non-trivial to specify manually in practice and thus rely more on the discovery from data. We will introduce how similar techniques are proposed for discovering various types of data dependencies.

Figure 1.4 compares the difficulties of data dependency discovery problems. Unfortunately, as illustrated, most problems are NP-complete. For instance, in CFD discovery, the problem of generating an optimal tableau for a given FD is NP-complete [88]. All the generalizations of CFDs, such as CDDs and DCs including CFDs as special cases in Fig. 1.1, inherit the difficulty, i.e., the discovery of CDDs and DCs is no easier than that of CFDs.

Application in Data Quality

Once the foundations of data dependencies are carefully constructed, understood and discovered, a practical issue is then how to apply such data dependencies in real data-centric

Fig. 1.4 The difficulties of data dependency discovery problems (assuming P≠NP and NP≠co-NP)

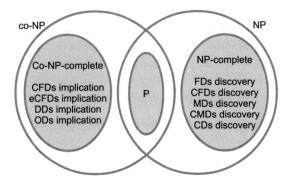

applications. Besides the traditional applications, such as schema design, integrity constraints, and query optimization with respect to the schema in databases, data dependencies are recently applied in data quality-oriented practice, e.g., violation detection [22], data repairing [44], consistent query answering [6], data deduplication [70], etc.

For instance, integrity constraints (e.g., FDs) are utilized to optimize the evaluation of queries, known as the *semantic query optimization* [33, 118]. Intuitively, various data dependencies can be exploited to rewrite the query, by eliminating redundant query expressions or using more selective access paths during query execution.

A more advanced form of data repairing is known as consistent query answering [6]. Given a set Σ of dependencies, a data instance I, and a query Q, consistent query answering aims to return certain answers to Q in I with respect to Σ, i.e., tuples in the answers to Q in each repair of I w.r.t. Σ, without editing I. When given FDs on identification data, PTIME bounds are developed by following a query rewriting approach [6]. However, it is still open if the problem considers other types of data dependencies, such as CFDs or MFDs.

Moreover, as indicated by Bohannon et al. [23], there exist strong connections between record linkage and data repairing. Specifically, repairs by data dependencies, e.g., FDs and INDs, which make inconsistent values identical, are similar to the process of record matching. As a step forward, the repairing task has to handle tuples across multiple tables, rather than to detect duplicate records in one or two relations.

In the chapter of each data type, we provide a table of the corresponding data dependencies supported for various application tasks. It is not surprising that a large number of studies are dedicated to the applications of violation detection and data repairing, referring to the veracity issue and the motivation of improving data quality. Moreover, the heterogeneous data type is well supported, as presented in Table 3.1, owing to the prevalent variety issue of big data. Application details of various data dependencies are discussed in the following sections.

1.5   Related Studies in Dependency Survey

We introduce below some related studies in dependency survey, e.g., by Caruccio et al. [31] on relaxed functional dependencies, Liu et al. [121] on dependency discovery algorithms, and so on.

As illustrated in Sect. 1.1, the traditional data dependencies such as FDs are declared on the whole relation, i.e., for any two tuples in a relation, if they have the same determinant X values, their dependent Y values must be equal as well. Such a strict condition cannot address the novel variety and veracity issues in big data. (1) Data dependencies might not hold in the whole relations. For example, one may notice that only in UK, zipcode determines street, but not in other countries. It leads to the extensions on statistics and conditions on categorical data in Chap. 2. (2) The same entity may have different representation formats in various data sources. For instance, "Chicago" and "Chicago, IL" denote the same region. To be tolerant to such heterogeneity, data dependencies are extended with distance metrics over heterogeneous data in Chap. 3. (3) The order of numerical data needs to be considered, such as the price increases on weekdays. Extensions are necessary to capture such order constraints over numerical data in Chap. 4. (4) As a special type of numerical data, we also study the constraints on timestamps over temporal data in Chap. 5. (5) Finally, the integrity constraints over graph data are summarized in Chap. 6.

Caruccio et al. [31] summarize some relaxations of the traditional functional dependencies. While the relaxed notations of various data dependencies are extensively introduced, it is not discussed on how data dependencies extend with each other, i.e., the family tree of extensions presented in Fig. 1.1. In addition to comparing FD with its more specific versions, in Fig. 1.3, we further compare the impact or importance of a data dependency, by counting the number of publications that use it. To add more insights into how data dependencies relate to each other, we provide a timeline of data dependencies that are proposed for various reasons, in Fig. 1.2. Important milestones are observed, such as AFDs [109] for data dependencies that approximately hold in a relation, and CFDs [22] on conditionally holding in a relation. Moreover, we compare the difficulties of data dependency discovery problems in Fig. 1.4. As illustrated, while most problems are NP-complete, the discovery problem for some data dependencies such as CSDs [87], however, is polynomial time solvable.

Liu et al. [121] review typical discovery algorithms for data dependencies mainly over categorical data, such as FDs, AFDs and CFDs. The discovery of data dependencies on heterogeneous and numerical data as well as their applications are not addressed.

Wang et al. [174] present a survey on accessing heterogeneous data, where some data dependencies are introduced for the application of semantic query optimization. However, only a limited number of data dependencies such as CDs are studied.

Abedjan et al. [2] provide a conceptual and technical overview of data profiling. Data dependencies are also included as multi-column profiling. Some classical data dependencies, such as FDs and INDs, are covered together with the corresponding discovery algorithms. Extensions of these dependencies such as CFDs are also introduced. Although both this work and ours discuss data dependencies, there are significant differences. Our work aims to outline the constraints on rich data types. We investigate data dependencies on various data types and summarize the extension relationships between these dependencies. Compared to [2], our work covers more constraints on richer data types, especially temporal and graph data.

1.6   Organization

In the rest of this study, we will introduce various data dependencies declared on different data types, including categorical, heterogeneous, numerical, temporal and graph data. Appendix A presents a full list of all the studied data dependencies, together with the corresponding section and page numbers. Table 1.2 lists the notations frequently used in this paper.

In Chap. 2, we first present the data dependencies on categorical data, since conventional data dependencies are often defined on the equality relationship of categorical data values. Extensions with statistics or conditions are made to meet the variety of big data. Chapter 3 introduces data dependencies over heterogeneous data. The data dependencies consider similarity metrics instead of equality operators, which increase expressive power. In Chap. 4, data dependencies on numerical data are discussed. Orders between two values are considered as the essential constraints in these dependencies. Chapter 5 introduces data dependencies on temporal data. Time series data become more and more popular in the IoT world and could be considered as an important data scenario. Although most of the

Table 1.2 Notations

Symbol	Description
R	Relation scheme
X, Y	Attribute sets in R
A, B	Single attributes in R
r	Relation instance
t	Tuple in r
t_p	Pattern tuple of conditions
$dom(A)$	Attribute domain of A
$\epsilon, \delta, \Delta, \alpha, \beta, g$	Threshold
Pr	Probability
θ, ϕ, μ	Similarity/differential/membership function
a, b	Weight
φ	Dependency

integrity constraints on relational data could also be used on temporal data, there are more time-related semantics to express the features of the temporal data better. In addition, we introduce data dependencies on graph data in Chap. 6. Traditional data dependencies cannot be directly used in graph data, since they are very different with relational data. Nevertheless, the idea of data dependencies could be extended to graphs. Finally, in Chap. 7, we conclude the study and discuss several promising directions for future work.

While categorical data, heterogeneous data, numerical data, temporal data, or even graph data are individually analyzed in each section, they would appear together in a system. Some data dependencies could express the constraints across different data types. As illustrated in Sect. 4.3, DCs can declare the constraints over both categorical and numerical data. For example, a DC may state that the price should not be lower than 200 (numerical value) in the region of "Chicago" (categorical value). Likewise, Sect. 3.4 shows that CDDs can express the constraints on both categorical and heterogeneous data. For instance, a CDD may state that in the region of "Chicago" (categorical value), two tuples (from heterogeneous sources) with similar name values (denoting the same hotel) should have similar address values. That is, while the data dependencies over categorical, heterogeneous and numerical data form three branches in Fig. 1.1, they do have connections. As aforesaid, DCs extend ODs for numerical data (Sect. 4.3) as well as eCFDs on categorical data (Sect. 4.3). Similarly, CDDs extend both DDs over heterogeneous data and CFDs over categorical data (Sect. 3.4).

Categorical Data

Data dependencies traditionally used for schema design [3] are often defined on the equality relation between categorical data values. For instance, a functional dependency (FD) $X \rightarrow A$ states that the same X values of two tuples imply their equal A values. Owing to the variety of issues in big data, data dependencies may not exactly hold. As indicated in [51, 139], the variety issue leads to constraints that may apply only to certain subsets of data. In this chapter, we introduce the extensions of data dependencies that are still based on equality relationships.

The categorical layer in Fig. 1.1 shows the data dependencies in this category, and also the corresponding extension relationships, i.e., where the data dependencies are extended from. Each pair of extensions (e.g., SFDs extend FDs) corresponds to an arrow from FDs to SFDs in the family tree in Fig. 1.1.

Table 2.1 further categorizes the extensions in the following aspects. (1) Statistical: A natural extension is to investigate the data dependencies that almost hold, known as statistical extensions. (2) Conditional: Another idea is to extend the data dependencies with conditions, i.e., data dependencies conditionally hold in a subset of data, namely conditional extensions. (3) Multivalued: Rather than X exactly determines Y, the multivalued dependencies state that X multidetermines Y, i.e., a particular X value is associated with a set of Y values and $R - XY$ values, and these two sets are independent of each other. (4) Inclusion: Finally, the inclusion dependencies extend the constraints between two relations, i.e., the constraints on existence.

© The Author(s), under exclusive license to Springer Nature Switzerland AG 2023   15
S. Song and L. Chen, *Integrity Constraints on Rich Data Types*,
Synthesis Lectures on Data Management,
https://doi.org/10.1007/978-3-031-27177-9_2

Table 2.1 Category and special cases of data dependencies over categorical data, where each pair corresponds to an arrow in Fig. 1.1 such as SFDs◁–FDs

Subcategory	Sections	Data dependencies	Special cases
Statistical	2.1	FDs	Keys
Statistical	2.2	EDGs	FDs
Statistical	2.3	SFDs	FDs
Statistical	2.4	PFDs	FDs
Statistical	2.5	AFDs	FDs
Statistical	2.6	NUDs	FDs
Conditional	2.7	CFDs	FDs
Conditional	2.8	eCFDs	CFDs
Multivalued	2.9	MVDs	FDs
Multivalued	2.11	AMVDs	MVDs
Multivalued	2.10	FHDs	MVDs
Inclusion	2.12	INDs	
Inclusion	2.13	AINDs	INDs
Inclusion	2.14	CINDs	INDs

2.1   Functional Dependencies (FDs)

Functional dependencies (FDs) have been long recognized as integrity constraints in databases [42]. They are first utilized in database design [3]. Conventionally, data dependencies are extracted from application requirements for database standardization and used to ensure data quality. For instance, functional dependencies (FDs) [16] are employed to evaluate whether a relation is in the third normal form (3NF) [42] or Boyce-Codd normal form (BCNF) [43]. FDs are extended to *multivalued dependencies* (MVDs) [13], i.e., every FD is also an MVD, in order to test whether a relation is in the fourth normal form (4NF) [55]. Such data dependencies are also used in database query optimization [53, 86].

Definition

A *functional dependency* (FD) is in the form of

$$FD : X \rightarrow Y,$$

where X, Y are attributes in a relation R. It states that for any two tuples from the instance of relation R, if they have the same X values, then their Y values must be equal as well.

Example

Consider a functional dependency in Table 1.1,

$$fd_1 : address \rightarrow region.$$

For example, tuples t_3 and t_4 have the same address value "#3, West Lake Rd." with an equal region value "New York". It means that each address corresponds to an explicit region.

Data quality issues of the relation instance $r_{1.1}$ can be detected by fd_1 in Table 1.1. For instance, it can show errors of region value in t_3 or t_4, i.e., "Boston" of t_3 is different from "Chicago, MA" of t_4, while they have the same address value.

Special Case: Keys

For a set of attributes K, the key dependency KEY(K) means no two tuples have the same K values, that is to say, K is a key [57]. Consider the relation instance $r_{1.1}$ in Table 1.1. The attribute name is the key of $r_{1.1}$. The key dependency KEY(name) can be represented as

$$fd : name \rightarrow address, region, star, price.$$

Axiomatization

For the class of categorical data, there exist a set of axioms (inference rules) for FDs, usually called Armstrong axioms [8].

The well-known Armstrong's axioms consist of three rules.

FD1 (Reflexivity) : If $Y \subseteq X$, then $X \rightarrow Y$.

FD2 (Augmentation) : If $X \rightarrow Y$, then $XZ \rightarrow YZ$.

FD3 (Transitivity) : If $X \rightarrow Y$ and $Y \rightarrow Z$, then $X \rightarrow Z$.

In the rules, X, Y, and Z are sets of attributes in a relation R.

Based on the set of rules, which is complete and sound, it is convenient to derive additional rules such as Union, Decomposition and Pseudo-transitivity [13].

FD4 (Decomposition) : If $X \rightarrow YZ$, then $X \rightarrow Y$ and $X \rightarrow Z$.

FD5 (Union) : If $X \rightarrow Y$ and $X \rightarrow Z$, then $X \rightarrow YZ$.

FD6 (Pseudo transitivity) : If $X \rightarrow Y$ and $YW \rightarrow Z$, then $XW \rightarrow Z$.

Since axiomatization reveals the insight of logical implication, it can be used for FD discovery, as discussed in Sect. 2.1 below.

Discovery

Data dependencies are often specified directly by the database designer or extracted from business requirements. That is, the semantics of data dependencies are studied before the data ingestion into databases. In contrast, in the big data era, rich data types are collected

Fig. 2.1 Searching space of
data dependency discovery

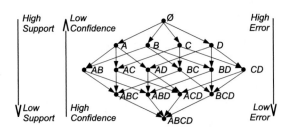

from existing sources with weakly specified integrity constraints. While manually specifying constraints is no longer an option over such big data, it often relies on automatically
discovering data dependencies from the data. The discovery algorithms for FDs are instruc-
tive, and many algorithms apply similar ideas to the discovery of other data dependencies.
Thus, we introduce the discovery algorithms for FDs in detail and provide some examples
below.

The discovery problem is to find the data dependencies that hold, or are expected to hold,
in a given dataset (either clean or dirty). For instance, given the instance in Table 1.1, it
is to find FDs such as fd_1 that (approximately) hold. The discovery of FDs from data is
known to be intrinsically hard, i.e., a minimal cover can be exponentially large with respect
to the number of attributes in a relation as shown in Fig. 2.1 [128, 129, 143]. The problem of
determining whether a relation has a key of size less than a given integer is NP-complete [12].
Efficient strategies are studied for discovering FDs, such as TANE [95, 96], FastFD [180].

FD discovery targets on generating a minimal cover of all FDs that hold in the given data.
According to the inference rules, Decomposition and Union, as presented at the beginning of
in this chapter, it is sufficient to discover dependencies in the embedded form of $X \rightarrow A$, i.e.,
with a single attribute in the right-hand side. For example, if two dependencies $fd_2 : B \rightarrow Y_1$
and $fd_3 : B \rightarrow Y_2$ hold, then $fd_3 : B \rightarrow Y_1 Y_2$ can be derived. Consequently, the discovery
targets on searching the space of possible X, i.e., searching the lattice of attributes as shown
in Fig. 2.1.

The level-wise algorithm, TANE [95, 96], searches data dependencies in the lattice of
attributes, together with efficient pruning. As shown in Fig. 2.1, TANE starts from the top
level, then moves downward level by level. In Level L_i, all attribute combinations $X \in L_i$
will be checked to see whether X can lead to an FD $X \backslash A \rightarrow A, A \in X$. If $X \backslash A \rightarrow A$
holds, all supersets of it will be pruned and will not be considered for generating attribute
combinations.

TANE discovers minimal FDs and prunes the search space following this idea. For
each attribute combination X, TANE maintains a set of right-hand side (RHS) candi-
dates $C^+(X) = \{A \in R \mid \forall B \in X : X \backslash \{A, B\} \rightarrow B \text{ does not hold}\}$. With $C^+(X)$, TANE
can ensure that there is no FD $X \backslash \{A, B\} \rightarrow B$ breaking the minimality of $X \backslash \{A\} \rightarrow B$.
If $C^+(X) = \emptyset$, TANE will prune the node X, because there is no candidate RHS attribute
for X.

For instance, assume we have a relation $R = \{A, B, C, D\}$ and minimal FDs $A \rightarrow B$, $BC \rightarrow D$ and $AC \rightarrow D$. By default, $L_0 = \emptyset$ and $C^+(\emptyset) = \{A, B, C, D\}$. TANE starts from $L_1 = \{\{A\}, \{B\}, \{C\}, \{D\}\}$. $C^+(A), C^+(B), C^+(C), C^+(D)$ are $\{A, B, C, D\}$, since $\emptyset \rightarrow Y, \forall Y \in R$, does not hold. Then, the algorithm moves to Level 2 by combining attributes, $L_2 = \{\{AB\}, \{AC\}, ..., \{CD\}\}$. $C^+(X)$ can be obtained as $C^+(X) = \bigcap_{X_i \in X} C^+(X)$. For example, $C^+(AB) = C^+(A) \cap C^+(B) = \{A, B, C, D\}$. For each attribute combination $X \in L_2$, we check whether $X \backslash X_i \rightarrow X_i$, $X_i \in X \cap C^+(X)$, is valid. If $X \backslash X_i \rightarrow X_i$ holds, then TANE will remove X_i and $X_j \in R \backslash X$ from $C^+(X)$. In this example, we find $A \rightarrow B$ after checking $A \rightarrow B$, $A \rightarrow C$, etc. And A, C, D are removed from $C^+(AB)$. In Level L_3, similar to L_2, we compute $C^+(X)$ first. For example, $C^+(ABC) = C^+(AB) \cap C^+(BC) \cap C^+(AC) = \{B\}$ and $C^+(BCD) = \{ABCD\}$. After checking, we find $BC \rightarrow D$ and $AC \rightarrow D$ hold and prune $C^+(BCD)$ and $C^+(ACD)$. In Level L_4, $C^+(ABCD) = \emptyset$, which means there is nothing left to check.

Remarkably, TANE algorithm also supports the discovery of approximate FDs that almost hold in the given data. In general, the higher the confidence of a data dependency is, the more likely it holds in the data. It also detects more accurately the errors in the violation detection application as introduced below.

In contrast, Flach and Savnik [79] discover FDs in a bottom-up style, which considers the maximal invalid dependencies first. When searching in a hypotheses space, the maximum invalid dependencies are used to prune the search space.

Application

FDs are utilized to optimize the evaluation of queries, known as *semantic query optimization* [33, 118]. Moreover, as indicated by Bohannon et al. [23], there is a close relationship between data repairing and record linkage. Especially, repairs by FDs, which make inconsistent values identical, are similar to the process of record matching. As a step forward, instead of detecting duplicate records in one or two relations, the repairing task has to process tuples across multiple tables.

2.2   Equality Generating Dependencies (EGDs)

Equality generating dependencies (EGDs) [15, 49] extend FDs by allowing equality relationships between multiple relations. EGDs say that if certain tuples satisfy certain patterns, some values of the tuple must be equal.

Definition

An *equality-generating dependency* (EGD) is in the form of:

$$\text{EGD} : \forall x_1 ... \forall x_n, \varphi(x_1, ..., x_n) \rightarrow \exists z_1 ... \exists z_k \psi(y_1, ..., y_m)$$

Table 2.2 An example relation instance $r_{2.2}$ of hotel

	Address	Region
t_1	No.5, Central Park	New York
t_2	No.5, Central Park	New York
t_3	#3, West Lake Rd.	Boston
t_5	Fifth Avenue, 61st Street	Chicago
t_7	No.7, West Lake Rd.	Boston, MA

where $\{z_1, ..., z_k\} = \{y_1, ..., y_m\} \backslash \{x_1, ..., x_n\}$. $\varphi(x_1, ..., x_n)$ is a conjunction of relational and equality atoms and ψ is a non-empty conjunction of equality atoms. An EGD states that if atoms in φ hold, then the equality relationships in ψ are valid.

Example

Consider the relation instance $r_{2.2}$ in Table 2.2. We have

$$\text{egd}_1 : \forall a \forall r \forall r' \text{hotel}(a, r) \wedge \text{hotel}(a, r') \rightarrow r = r'.$$

This egd_1 means that if two hotels have the same address, then their regions should be equal. For instance, as shown in Table 2.2, t_1 and t_2 have the same address ("No.5, Central Park"), then their regions should be the same ("New York").

Special Case: FDs

As indicated in Fig. 1.1, EGDs extend FDs. According to Deutsch [49], any FD can be expressed as the EGD:

$$\forall x_1 \ldots \forall x_n, \forall y_1 \ldots \forall y_m, \forall y'_1 \ldots \forall y'_m, \forall z_1 \ldots \forall z_k, \forall z'_1 \ldots \forall z'_k,$$
$$R(x_1, \ldots, x_n, y_1, \ldots, y_m, z_1, \ldots, z_k) \wedge R(x_1, \ldots, x_n, y'_1, \ldots, y'_m, z'_1, \ldots, z'_k)$$
$$\rightarrow y_1 = y'_1 \wedge \ldots \wedge y_m = y'_m$$

where X and Y are lists of attributes in the relation $R = [X, Y, Z]$, $Z = R \backslash X \bigcup Y$, $|X| = n$, $|Y| = m$, $|Z| = k$. For example, the fd_1 : attributes $\rightarrow$ region over $r_{2.2}$ can be expressed as egd_1.

Application

EGDs can capture equality relationships between attributes from multiple relations. Thus, Wang [168] proposes to apply EGDs to optimize the database design by detecting and removing redundancies across multiple relations. In ontology querying, $\text{Datalog}^{\pm}$ family, a prominent family of languages, expresses the ontological theory with EGDs [30]. In addition, Calautti et al. [29] exploit EGDs in checking chase termination.

2.3   Soft Functional Dependencies (SFDs)

In real world, some data values determine other values not with certainty but merely with high probability. It is not practical to use hard FDs to detect this relationship. For this reason, Ilyas et al. [98] study a notation of soft FDs (SFDs), where the values of an attribute are well-predicted by the values of another. It is a generalization of the classical notion of hard FDs where the value of X completely determines that of Y. For example, in a database of cars, a soft functional dependency could be: model $\rightarrow$ make. Given that model = 323, we know that make = Mazda with high probability, but there is also a small chance that make = BMW. Such soft FDs are useful in improving selectivity estimation during query optimization [98] and recommending secondary indices [107].

Definition

A *soft functional dependency* (SFD) is in the form of

$$SFD : X \rightarrow_s Y,$$

where X, Y are attributes in a relation R, and s is a minimum threshold of strength measure. The strength measure quantifies at what level $X \rightarrow Y$ holds in a relation instance r

$$S(X \rightarrow Y, r) = \frac{|dom(X)|_r}{|dom(X, Y)|_r},$$

where $|dom(X)|_r$ is the number of distinct values in attributes X in r, and $|dom(X, Y)|_r$ is the number of distinct values in the concatenation of X and Y in r. A SFD has $S(X \rightarrow Y, r) \geq s$, stating that the value of X determines that of Y not with certainty, but with high probability.

Example

Consider the relation instance $r_{2.3}$ in Table 2.3. We have

$$S(\text{address} \rightarrow \text{region}, r_{2.3}) = \frac{|dom(X)|_{r_{2.3}}}{|dom(X, Y)|_{r_{2.3}}} = \frac{2}{3}.$$

Table 2.3 An example relation instance $r_{2.3}$ of hotel where FD address $\rightarrow$ region almost holds, while name $\rightarrow$ address is not clear to hold

	Name	Address	Region	Rate
t_1	Hyatt	175 North Jackson Street	Jackson	230
t_2	Hyatt	175 North Jackson Street	Jackson	250
t_3	Hyatt	6030 Gateway Boulevard E	El Paso	189
t_4	Hyatt	6030 Gateway Boulevard E	El Paso, TX	189

That is, address $\rightarrow$ region almost holds in $r_{2.3}$. Instead, the strength measure for name $\rightarrow$ address in relation $r_{2.3}$ is

$$S(\text{name} \rightarrow \text{address}, r_{2.3}) = \frac{|dom(X)|_{r_{2.3}}}{|dom(X, Y)|_{r_{2.3}}} = \frac{1}{2}.$$

It is not a clear FD with lower strength value.

Special Case: FDs

As indicated in Fig. 1.1, SFDs extend FDs. When the value of X determines the value of Y with strength 1, it is exactly an FD. In other words, all FDs can be represented as special SFDs with $s = 1$. For example, fd_1 in Sect. 1.1 can be equivalently represented by

$$sfd_1 : \text{address} \rightarrow_1 \text{region}.$$

That is, for the relation $r_{1.1}$ in Table 1.1, we have strength $S(\text{address} \rightarrow \text{region}, r_{1.1}) = 1$. In this sense, SFDs subsume the semantics of FDs, or SFDs generalize/extend FDs, denoted by the arrow from FDs to SFDs in Fig. 1.1.

Discovery

Ilyas et al. [98] propose a sample-based approach CORDS to discover SFDs, which uses system catalog to retrieve the number of distinct values of a column. It uses a robust chi-square analysis to identify the correlation between attributes, and analyzes the number of different values in the sampling column to detect SFDs. The sample size of the algorithm is basically independent of the database size. Thereby, the algorithm is a data-driven approach [97] and highly scalable. Kimura et al. [107] describe algorithms to search for SFDs that can be exploited at query execution time by introducing appropriate predicates and choosing a different index. It introduces buckets on the domains of both attributes to reduce the size of index.

Application

Instead of strict constraints, soft functional dependencies are useful in improving the selectivity estimation during query optimization [98] by collecting joint statistics for those correlated data columns. Specifically, SFDs exploit correlated attributes as additional predicates to queries. Then, a clustered index on such additional attributes could be utilized for query optimization.

SFDs also can be used for improving query processing performance [107], owing to the property that the values of an attribute are well-predicted by the values of some others. If a column is correlated to another column, it is possible to recommend secondary indices.

2.4   Probabilistic Functional Dependencies (PFDs)

Many applications need to extract structured data from the Web. In traditional database design, FDs are designated as the true statement of all possible instances of the database. However, in the Web environment, FDs are often discovered by counting-based algorithm on multiple data sources. In this sense, FDs need to be expanded with probability to capture the uncertainty. Therefore, Wang et al. [167] extend functional dependencies with probability for data integration systems, namely probabilistic functional dependencies (PFDs). This extension of traditional FDs is still on statistics (counting) of equal values.

Definition

A *probabilistic functional dependency* (PFD) over attributes X and Y in relation R is denoted by

$$\text{PFD} : X \to_p Y,$$

where $X \to Y$ is a standard FD, and p is a maximum threshold of the likelihood that the FD $X \to Y$ is correct.

To compute the likelihood in a relation instance r, it first calculates the fraction of tuples for each distinct value V_X of X. Let V_Y be the Y-value that occurs in the maximum number of tuples with value V_X in X. The probability is

$$P(X \to Y, V_X) = \frac{|V_Y, V_X|}{|V_X|},$$

where $|V_Y, V_X|$ is the number of tuples with values V_X for X and V_Y for Y, and $|V_X|$ is the number of tuples with values V_X for X. The probability of $X \to Y$ in r is given by the average of probabilities for each distinct value of X,

$$P(X \to Y, r) = \frac{\sum_{V_X \in D_X} P(X \to Y, V_X)}{|D_X|},$$

where D_X is all distinct values of X in r. A PFD has $P(X \to Y, r) \geq p$, i.e., a high probability to hold.

Example

Consider again the relation instance $r_{2.3}$ in Table 2.3, where an FD address $\to$ region almost holds. We have

$$P(\text{address} \to \text{region}, V_1) = 1,$$

$$P(\text{address} \to \text{region}, V_2) = \frac{1}{2},$$

$$P(\text{address} \to \text{region}, r_{2.3}) = \frac{3}{4},$$

given $V_1 =$ "175 North Jackson Street" and $V_2 =$ "6030 Gateway Boulevard E". Then, there is a PFD

$$\mathsf{pfd}_1 : \mathsf{address} \to_{3/4} \mathsf{region}.$$

Similarly, for the FD $\mathsf{name} \to \mathsf{address}$, which does not clearly hold in Table 2.3, we have

$$P(\mathsf{name} \to \mathsf{address}, V_1) = \frac{|V_{\mathsf{address}}, V_1|}{|V_1|} = \frac{1}{2},$$

where $V_1 =$ "Hyatt", $V_{\mathsf{address}} =$ "6030 Gateway Boulevard E". It follows

$$P(\mathsf{name} \to \mathsf{address}, r_{2.3}) = \frac{1}{2}.$$

The corresponding PFD can be denoted as

$$\mathsf{pfd}_2 : \mathsf{name} \to_{1/2} \mathsf{address}.$$

Special Case: FDs

From Fig. 1.1, we can see that PFDs subsume FDs. When the value of X determines the value of Y with $P(X \to Y, r) = 1$, having $p = 1$ in a PFD $X \to_p Y$, it is exactly an FD. That is, all FDs can be represented as special PFDs with $p = 1$. For example, fd_1 in Sect. 1.1 can be represented by

$$\mathsf{pfd}_3 : \mathsf{address} \to_1 \mathsf{region}.$$

That is, for the relation $r_{1.1}$ in Table 1.1, we have $P(\mathsf{address} \to \mathsf{region}, r_{1.1}) = 1$. Consequently, PFDs subsume the semantics of FDs, or PFDs generalize/extend FDs, denoted by the arrow from FDs to PFDs in Fig. 1.1.

Discovery

Wang et al. [167] extend the TANE algorithm over a single table to generate PFDs from hundreds of small, dirty and incomplete data sets. Two counting-based algorithms are proposed. The first algorithm merges the values and computes the probability of each FD, while the second algorithm, which is for multiple sources, merges PFDs obtained from each source.

Application

Given a mediated schema and some mappings from each source to the mediated schema, the probability of an FD in each data source is merged together as a global measure. Wang et al. [167] propose to use PFDs to solve two problems arose in pay-as-you-go systems, in order to gauge and improve the quality of the information integration.

First, FDs are declared with probabilities to capture the inherent uncertainties over many data sources. The violation of PFDs by some data sources can help pinpoint data sources with low-quality data. Moreover, PFDs can also help normalize a large automatically generated mediated schema into relations that correspond to meaningful real-world entities and relationships, to help users better understand the underlying data.

2.5   Approximate Functional Dependencies (AFDs)

Rather than FDs exactly holding, approximate functional dependencies (AFDs) [109] declare FDs that almost hold in a relation.

AFDs have similar semantics with SFDs, but with a different statistical measure. As illustrated in Sect. 2.3, the strength measure may not precisely reflect whether an FD almost holds.

Definition

An *approximate functional dependency* (AFD) between attributes X and Y in R is denoted by

$$\text{AFD} : X \rightarrow_{\varepsilon} Y,$$

where ε is a maximum threshold of an error measure evaluating the exact proportion of tuples with violations.

To evaluate the proportion of tuples with violations to an FD, the g_3 error measure [108] is employed. Given a relation instance r, the g_3 error measure calculates the ratio of the minimum number of tuples that need to be removed from r to make $X \rightarrow Y$ hold on r

$$g_3(X \rightarrow Y, r) = \frac{|r| - \max\{|s| \mid s \subseteq r, s \vDash X \rightarrow Y\}}{|r|},$$

where s is a subset of tuples in r that do not violate $X \rightarrow Y$, denoted by $s \vDash X \rightarrow Y$. The measure g_3 defines the approximation of a dependency $X \rightarrow Y$. A natural interpretation is the fraction of rows with exceptions or errors affecting the dependency. The smaller the g_3 measure is, the more likely to be an FD.

An AFD has $g_3(X \rightarrow Y, r) \leq \varepsilon$, where the error threshold ε has $0 \leq \varepsilon \leq 1$. $X \rightarrow Y$ is an approximate dependency discovered from a relation r, if and only if $g_3(X \rightarrow Y, r)$ is at most ε.

Example

Consider again the relation instance $r_{2.3}$ in Table 2.3. We have

$$g_3(\text{address} \rightarrow \text{region}, r_{2.3}) = \frac{1}{4}.$$

By removing either t_3 or t_4, the violation eliminates. It can be computed by grouping tuples according to equal X values, i.e., on address, then finding the minimum violation tuples in each group.

Similarly, for name $\rightarrow$ address in $r_{2.3}$, we have

$$g_3(\text{name} \rightarrow \text{address}, r_{2.3}) = \frac{1}{2}.$$

That is, at least two tuples need to be removed (e.g., t_3 and t_4) in order to make the FD hold. Data dependencies address $\rightarrow$ region are more likely to hold in $r_{2.3}$ with a lower g_3 error measure.

Special Case: FDs

Again, in Fig. 1.1, we show that AFDs generalize FDs. For an FD, i.e., the value of X determines exactly that of Y, the corresponding error measure is $g_3(X \rightarrow Y, r) = 0$. The example fd_1 in Sect. 1.1 can be equivalently represented by

$$\text{afd}_1 : \text{address} \rightarrow_0 \text{region}.$$

In other words, for the relation $r_{1.1}$ in Table 1.1, we have error measure $g_3(\text{address} \rightarrow \text{region}, r_{1.1}) = 0$. Thereby, AFDs subsume the semantics of FDs, or AFDs generalize/extend FDs, denoted by the arrow from FDs to AFDs in Fig. 1.1.

Discovery

The TANE algorithm [95, 96] for discovering exact FDs can also be adapted for AFDs discovery.

Similar to TANE, it partitions the set of rows based on attribute values for handling a large number of tuples. The use of partitions also makes the discovery of AFDs simple and efficient. The error or abnormal tuples can be easily identified. The key modification is to change the validity test on whether $X \rightarrow Y$ exactly holds to whether $g_3(X \rightarrow Y, r) \leq \varepsilon$ as defined in Sect. 2.5. It computes all minimal approximate dependencies $X \rightarrow Y$ with $g_3(X \rightarrow Y, r) \leq \varepsilon$, for a given threshold value ε.

Application

AFDs are utilized for query processing over incomplete databases. To retrieve possible answers, the QPIAD system [179] mines the inherent correlations among database attributes represented as AFDs. These AFDs are exploited to select features and compute probability distribution over the possible values of the missing attribute for a given tuple.

2.6   Numerical Dependencies (NUDs)

One-to-many relationships are common in practice, e.g., the same object may have different names under various naming conventions. In this sense, FDs are too strict, requiring that an element of a particular attribute or set of attributes can be associated with only one element.

For this concern, Kivinen and Mannila [109] relax FDs and propose numerical dependencies (NUDs). NUDs state that with an element of a particular attribute or set of attributes, one can associate up to k elements of another attribute or set of attributes.

Definition

A *numerical dependency* (NUD) on a relation R has the form

$$\text{NUD} : X \rightarrow_k Y,$$

where X, Y are attribute sets in R, and $k \geq 1$ is called the weight of the NUD. It states that each value of X can never be associated to more than k distinct values of Y.

Example

Consider an NUD over relation $r_{2.3}$ in Table 2.3,

$$\text{nud}_1 : \text{address} \rightarrow_2 \text{region}.$$

It states that one address can only have at most 2 variations of region. As shown in Table 2.3, there are 2 different region representation formats for "El Paso" in tuples t_3 and t_4.

Special Case: FDs

Figure 1.1 shows that FDs are special cases of NUDs. All FDs can be represented as special NUDs with $k = 1$. For example, fd_1 in Sect. 1.1 can be equivalently represented by

$$\text{nud}_2 : \text{address} \rightarrow_1 \text{region}.$$

That is, for the relation $r_{1.1}$ in Table 1.1, each tuple $t[X]$ is associated to at most one value on Y. Thus, NUDs subsume the semantics of FDs, or NUDs generalize/extend FDs, denoted by the arrow from FDs to NUDs in Fig. 1.1.

Application

NUDs can be used in various scenarios [41] such as (1) estimating the projection size of a relation, that is, number of distinct attribute values of a subset; (2) estimating the cardinality of aggregate views; and (3) efficient query processing in nondeterministic databases.

2.7 Conditional Functional Dependencies (CFDs)

The conditional functional dependencies (CFDs), as an extension of traditional FDs with conditions, are proposed in [22, 65] for data cleaning. The basic idea of CFDs is making the FDs, originally holding for the whole table, valid only for a set of tuples specified by the conditions. FDs and statistical extensions are often inadequate to capture sufficient semantics of data. Unlike traditional FDs, CFDs aim to obtain data consistency by combining binding of semantically related values. In other words, rather than FDs that hold in the whole relation, conditional functional dependencies (CFDs) [22, 65] declare FDs that conditionally hold in a part of the relation.

Definition

A *conditional functional dependency* (CFD) over R is a pair

$$\text{CFD} : X \rightarrow Y, t_p,$$

where (1) X and Y are attributes in R; (2) $X \rightarrow Y$ is a standard FD, embedded in CFD; and (3) t_p is a pattern tuple with attributes in X and Y. For each $B \in X \cup Y$, $t_p[B]$ is either a constant 'a' in $dom(B)$, or an unnamed variable '_' that draws values from $dom(B)$. It denotes that $X \rightarrow Y$ conditionally holds over a subset of tuples specified by t_p.

Example

A CFD over the relation instance $r_{2.3}$ in Table 2.3 can be

$$\text{cfd}_1 : \text{region, name} \rightarrow \text{address}, (\text{Jackson}, \_ \parallel \_).$$

It assures that for the tuples whose region is "Jackson", if they have the same name, then their address value must be equal (since there is only one Hyatt hotel in Jackson). For better readability, the CFD can be also written as

$$\text{cfd}_1 : \text{region} = \text{"Jackson", name} = \_ \rightarrow \text{address} = \_.$$

Tuples t_1 and t_2 satisfy cfd_1, which have the same name, the same region of "Jackson", as well as the same address.

Special Case: FDs

CFDs extend FDs as shown in Fig. 1.1. When the value of X determines the value of Y without conditions, it is exactly an FD. In other words, all FDs can be represented as special CFDs without constants in t_p. The example fd_1 in Sect. 1.1 can be represented as a CFD,

$$\text{cfd}_2 : \text{address} \rightarrow \text{region}, (\_ \parallel \_).$$

For better readability, it can be also written as

$$\text{cfd}_2 : \text{address} = \_ \rightarrow \text{region} = \_.$$

Therefore, CFDs subsume the semantics of FDs, or CFDs generalize/extend FDs, denoted by the arrow from FDs to CFDs in Fig. 1.1.

Axiomatization

Analogous to Armstrong's axioms, an inference system is provided for CFDs [65]. There are four axioms in the axiomatization for CFDs,

CFD1 : If $A \in X$, then $\left(R : X \to A, t_p\right)$, where $t_p[A_L] = t_p[A_R] = $ 'a' for some 'a' $\in \text{dom}(A)$, or both are equal to '_'.

CFD2 :   If (1) $(R : X \to A_i, t_i)$ such that $t_i[X] = t_j[X]$ for all $i, j \in [1, k]$, (2)$(R : [A_1, \ldots, A_k] \to B, t_p)$, and (3) $(t_1[A_1], \ldots, t_k[A_k]) \preceq t_p[A_1, \ldots, A_k]$, then $\left(R : X \to B, t_p'\right)$, where $t_p'[X] = t_1[X]$ and $t_p'[B] = t_p[B]$.

CFD3 :   If $\left(R : [B, X] \to A, t_p\right), t_p[B] = $ '_', and $t_p[A]$ is a constant, then $\left(R : X \to A, t_p'\right)$, where $t_p'[X \cup \{A\}] = t_p[X \cup \{A\}]$.

CFD4 :   If (1) $\Sigma \vdash_I (R : [X, B] \to A, t_i)$ for $i \in [1, k]$, (2) $\text{dom}(B) = \{b_1, \ldots, b_k, b_{k+1}, \ldots, b_m\}$, and $(\Sigma, B = b_l)$ is not consistent except for $l \in [1, k]$, and (3) for $i, j \in [1, k], t_i[X] = t_j[X]$ and $t_i[B] = b_i$, then $\Sigma \vdash_I \left(R : [X, B] \to A, t_p\right)$, where $t_p[B] = $ '_' and $t_p[X] = t_1[X]$.

The first two inference rules extend Armstrong's axioms of Reflexivity and Augmentation. Since FDs have no pattern tuples, the other two axioms for CFDs cannot find a counterpart in Armstrong's axioms. The third axiom considers the implication about constant and wildcard in the conditions. And the fourth inference rule for CFDs deals with attributes of finite domains.

Discovery

Note that for discovering data dependencies with conditions, it has to find not only the attribute sets but also the condition patterns on the attribute sets. When a rule of attribute sets $X \to Y$ is suggested, Golab et al. [88] study the discovery of optimal CFDs with the minimum pattern tableau size. A concise set of patterns are naturally desirable which may have lower cost during the applications such as violation detection by CFDs. For CFDs discovery, the problem of generating an optimal tableau for a given FD is NP-complete [88]. The implication problem for CFDs is co-NP-complete [22]. Since CFDs hold only in a subset of tuples rather than the entire table as FDs, an important problem in CFD discovery is thus to evaluate how many tuples the discovered constraints can cover, known as the support of a CFD.

When the embedded FDs are not given, Fan et al. [66, 67] propose three algorithms to discover CFDs. CFDMiner, based on the connection between minimal constant CFDs, finds constant CFDs by leveraging mining technique. CTANE extends TANE [95, 96] to discover general minimal CFDs, based on attribute-set/pattern tuple lattice. Wyss et al. [180] study depth-first, heuristic-driven algorithm, namely FastFDs, which is (almost) linear to the size of FDs cover. FastCFD, an extension of FastFD [180], discovers general CFDs by employing a depth-first search strategy. A level-wise algorithm, proposed by Chiang and Miller [36], uses an attribute lattice to generate candidate embedded FDs. Yeh et al. [181] present Data Quality Rules Accelerator (DQRA), by gradually introducing additional conditions to some initial CFDs. DQRA first generates all attribute pair combinations and candidate CFDs by pruning useless pairs based on the correlation between attributes. DQRA then revises inconsistencies in the initial CFDs by constraining its antecedent with additional conditions. Finally, it filters weak CFDs using the measure of conviction, and generalizes all discovered CFDs to get the final results.

Application

Informally, given a set of data dependencies, the violation detection problem is to find those tuples that do not satisfy the given dependencies, i.e., violations to the constraints. In the detection of CFDs violations [22], the main challenge comes from a large number of data values as conditions in data dependencies, which is different from traditional FDs. Since identification data are considered, an efficient SQL implementation is developed for detecting CFD violations.

Cong et al. [44] study the detecting and repairing methods of violations to CFDs. Two strategies are investigated to improve data consistency, or (1) directly computing a repair that satisfies a given set of CFDs, (2) incrementally finding a repair with updates to a database. Due to the hardness of repair problems, heuristic algorithms are developed as well. Given a set Σ of dependencies and an instance I of a schema R, data repairing is to find a repair I' of I such that I' satisfies Σ and has a minimal difference from I. An inconsistent data instance could be repaired via tuple insertion, deletion, or value modification (see [58] for a survey). The repairing by using CFDs is also studied by Cong et al. [44], which can capture more inconsistencies than what standard FDs can detect. As the hardness of FDs repairing, the problem of finding repair for CFDs remains NP-complete.

Fan et al. [75] investigate the propagation problem of CFDs. Given a set of CFDs on a data source, it is to determine whether or not such CFDs are still valid on the views (mappings) of the given data source. Such propagation is useful for data integration, data exchange and data cleaning. Algorithms are developed for computing the cover of CFDs that are propagated from the original source to the views. As novel types of integrity constraints, CFDs are naturally applicable in this traditional application of writing queries.

2.8 Extended Conditional Functional Dependencies (eCFDs)

Further extensions on conditional functional dependencies have been drawn as well, to further capture the more complicated semantics.

To substantially improve the expressive power of CFDs, Bravo et al. [25] and Ma et al. [125] study an extension of CFDs by employing disjunction and specifying the constant patterns of data with $\neq, <, \leq, >$ and $\geq$ predicates instead of only equality $=$, known as extended conditional functional dependencies (eCFDs). Specifically, eCFDs specify the pattern of semantically related values using disjunction and inequality, which cannot be expressed by CFDs. Similar extensions are also proposed by Golab et al. [88], which define a range tableau for CFDs, where each value is a range.

Definition

An *extended conditional functional dependency* (eCFD) is

$$\text{eCFD} : X \rightarrow Y, t_p,$$

where (1) X and Y are attributes in R; (2) $X \rightarrow Y$ is an embedded standard FD; and (3) t_p is a pattern tuple with attributes in X and Y. For each A in $X \cup Y$, $t_p[A]$ is either an unnamed variable '_' that draws values from $dom(A)$, or 'op a', where op is one of $\{=, \neq, <, \leq, >, \geq\}$, and '$a$' is a constant in $dom(A)$. That is, more operators are employed to specify the subset of tuples where the embedded FD holds.

Example

Again, we provide an example about eCFDs as follows. An eCFD over the relation instance $r_{2.3}$ in Table 2.3 can be

$$\text{ecfd}_1 : \text{rate}, \text{name} \rightarrow \text{address}, (\leq 200, \_ \parallel \_),$$

which can also be written as follows:

$$\text{ecfd}_1 : \text{rate} \leq 200, \text{name} = \_ \rightarrow \text{address} = \_.$$

It states that if two tuples such as t_3 and t_4 in Table 2.3 have the same rate value ≤ 200, then their name determines address (since small cities often have one hotel of each brand with a relatively low rate).

Special Case: CFDs

We show in Fig. 1.1 that eCFDs extend CFDs in Sect. 2.7. When only the $=$ operator is declared in an eCFD, it is exactly a CFD. In other words, all CFDs can be represented as special eCFDs. Indeed, the example cfd$_1$ in Sect. 2.7 is naturally an eCFD with operator $=$ only

$$\text{ecfd}_2 : \text{region} = \text{``Jackson''}, \text{name} = \_ \rightarrow \text{address} = \_.$$

In this sense, eCFDs subsume the semantics of CFDs, or eCFDs generalize/extend CFDs, denoted by the arrow from CFDs to eCFDs in Fig. 1.1.

There are some differences in terms of extension between CFDs and eCFDs. That is, as illustrated in Fig. 1.1, CDDs extend CFDs but not eCFDs.

Discovery

Zanzi and Trombetta [183] propose an algorithm with complexity of $O(n^2)$ with n as the size of a relation, for discovering non-constant eCFDs from a dataset. The discovered eCFDs have only one right-hand side attribute, and the conditions with operators in addition to $=$ are considered only for numerical attributes. The algorithm finds tuples which cannot satisfy a target dependency, and uses these tuple values to establish conditions for valid dependencies.

It uses small-to-large search to generate candidates for target dependencies and reduces the number of generated candidates by some pruning methods. Then, the determined candidates are verified whether they can be eCFDs and the values for attributes are determined, which have to be excluded for a valid eCFD. In other words, in order to obtain eCFDs, this algorithm distinguishes the excluding set and accepting set.

The complexity of the implication problem for eCFDs, implying an eCFDs from a set of eCFDs, remains unchanged, i.e., coNP-complete as CFDs [25].

Application

The eCFDs can be used to detect data inconsistencies, impose constraints on data and represent relationships among data. Bravo et al. [25] propose a batch algorithm to find all tuples that violate eCFDs. A single pair of SQL queries can find all violations of eCFDs. In addition, an incremental algorithm is developed. Specifically, an approximation algorithm MAXGSAT was presented to find an approximation of the largest possible subset of satisfiable eCFDs in a given input set D. This not only allows to efficiently detect the unsatisfiability of D with some certainty, but also provides the user with a set of satisfiable eCFDs that can serve as a starting point to inspect the remaining eCFDs in D. For the updates Δ_D to the database D, the algorithm can find all eCFDs violations with a good performance in the updated $D \oplus \Delta_D$, since this algorithm can update eCFDs violations in D and make the unnecessary re-computation minimized.

2.9   Multivalued Dependencies (MVDs)

In relational schema, FDs cannot represent the one-to-many relationships between attribute values. For FDs, to check whether there is a relationship between X and Y, we only need to investigate the two groups of properties of X and Y, which have nothing to do with other attributes. However, in real data, although some attributes have no direct relationship, there are indirect relationships between them. In multivalued dependencies (MVDs) [13, 55], whether X and Y have multivalued dependency depends on attribute Z.

While functional dependencies (FDs), a subclass of equality-generating dependencies, as mentioned in Sect. 2.2, rule out the existence of certain tuples, multivalued dependencies (MVDs), as a special form of tuple-generating dependencies, require the presence of certain tuples.

Definition

A *multivalued dependency* (MVD) over relation R is in the form of

$$\text{MVD} : X \twoheadrightarrow Y,$$

where $X \cup Y \cup Z = R$ is a partition of R. A relation instance r of R satisfies the MVD, if a given pair of (X, Z) values has a set of Y values, which are determined only by X values and independent of Z values, i.e., $r = \pi_{XY}(r) \bowtie \pi_{XZ}(r)$.

Example

An MVD in relation $r_{2.3}$ in Table 2.3 can be

$$\text{mvd}_1 : \text{address}, \text{rate} \twoheadrightarrow \text{region},$$

where $X = \{\text{address}, \text{rate}\}$, $Y = \{\text{region}\}$, $Z = \{\text{name}\}$. As shown in Tables 2.4 and 2.5, we have lossless decomposition, i.e., $r_{2.3} = \pi_{XY}(r_{2.3}) \bowtie \pi_{XZ}(r_{2.3})$. For instance, given the (address, rate, name) values "(6030 Gateway Boulevard E, 189, Hyatt)" in t_3 and t_4, we have a set of region values {"El Paso", "El Paso, TX"} independent of name. Similarly, for t_1 with (address, rate, name) values (175 North Jackson Street, 230, Hyatt), we also have region value "Jackson" independent of name.

Special Case: FDs

From Fig. 1.1, we can see that FDs are special cases of MVDs. All FDs can be represented as special MVDs. For example, fd_1 in Sect. 1.1 can be rewritten as

$$\text{mvd}_2 : \text{address} \twoheadrightarrow \text{region}.$$

That is, the value on address determines the set of values on region, having set size 1. Thereby, MVDs subsume the semantics of FDs, or MVDs generalize/extend FDs, denoted by the arrow from FDs to MVDs in Fig. 1.1.

Table 2.4 Decomposed XY of $r_{2.3}$ where $X = \{\text{address}, \text{rate}\}$, $Y = \{\text{region}\}$

	Address	Region	Rate
t_1	175 North Jackson Street	Jackson	230
t_2	175 North Jackson Street	Jackson	250
t_3	6030 Gateway Boulevard E	El Paso	189
t_4	6030 Gateway Boulevard E	El Paso, TX	189

Table 2.5 Decomposed XZ of $r_{2.3}$ where $X = \{\text{address}, \text{rate}\}$, $Z = \{\text{name}\}$

	Name	Address	Rate
t_1	Hyatt	175 North Jackson Street	230
t_2	Hyatt	175 North Jackson Street	250
t_3	Hyatt	6030 Gateway Boulevard E	189

Axiomatization

Four rules constitute a complete set of inference rules for MVDs [13], including Complementation, Reflexivity, Augmentation and Transitivity.

MVD1 (Complementation): Let X, Y, Z be sets of attributes, $U = X \cup Y \cup Z$ and $Y \cap Z \subseteq X$. Then $X \twoheadrightarrow Y$ if and only if $X \twoheadrightarrow Z$.

MVD2 (Reflexivity): If $Y \subseteq X$, then $X \twoheadrightarrow Y$.

MVD3 (Augmentation): If $Z \subseteq W$ and $X \twoheadrightarrow Y$, then $XW \twoheadrightarrow YZ$.

MVD4 (Transitivity): If $X \twoheadrightarrow Y$ and $Y \twoheadrightarrow Z$, then $X \twoheadrightarrow Z - Y$.

MVD2, MVD3, MVD4 extend Armstrong's axioms. MVD1 cannot find a counterpart in Armstrong's axioms and is usually applied when Z is the complement (in U) of either Y or the union of X and Y.

Discovery

Savnik and Flach [142] study the discovery of MVDs from relations. According to the generalization relationships between MVDs, it searches for valid MVDs in the hypothesis space designed. The idea is generally similar to the level-wise search of FDs discovery [95, 96]. Two strategies for discovering MVDs are proposed. The top-down algorithm searches for the positive border of valid dependencies, from most general dependencies to more specific ones. The bottom-up algorithm first calculates the negative border of invalid MVDs by eliciting false dependencies.

Application

While MVDs are extremely important in database design, with the fourth normal form (the original relation can be decomposed by MVDs and obtained from the new relations by taking joins) [55], Salimi et al. [140] recently introduce a novel application of MVDs to guarantee model fairness in machine learning. The training data often reflects discrimination, e.g., on race or gender, which is difficult to eliminate owing to the causal relationships among attributes. Intuitively, MVDs can be employed to capture the conditional independence. The fairness is thus reduced to a database repair problem by linking causal inference to multivalued dependencies (MVDs).

2.10   Full Hierarchical Dependencies (FHDs)

While multivalued dependencies (MVDs) decompose a relation into two of its projections without loss of information, the full hierarchical dependencies (FHDs) [48, 92] further study the hierarchical decomposition of a relation into multiple relations.

FHDs, interpreted as a set of MVDs, construct a large class of relational dependencies including MVDs, i.e., FHDs generalize MVDs. As mentioned in [92], an extension of the subset rule from MVDs to FHDs plays a key role in achieving complementarity.

Definition

A *full hierarchical dependency* (FHD) is an expression

$$\text{FHD} : X : \{Y_1, \ldots, Y_k\},$$

where $X, Y_1, \ldots, Y_k \subseteq R$ form a partition of relation R. A relation instance r of R satisfies the FHD, if $r = \pi_{XY_1}(r) \bowtie \cdots \bowtie \pi_{XY_k}(r) \bowtie \pi_{X(R-XY_1\ldots Y_k)}(r)$. That is, the FHD decomposes r into multiple new relations without loss of information. When $k = 1$, it is exactly an MVD $X \twoheadrightarrow Y_1$.

Example

Take relation $r_{2.6}$ in Table 2.6 as an example. The FHD is as follows:

$$\text{fhd}_1 : \text{Title} : \{\text{Actor}, \text{Feature}\}.$$

Two MVDs are given in the FHD, i.e.,

$$\text{mvd}_3 : \text{Title} \twoheadrightarrow \text{Actor},$$

$$\text{mvd}_4 : \text{Title} \twoheadrightarrow \text{Feature}.$$

Consequently, Table 2.6 is decomposed into Tables 2.7, 2.8 and 2.9. The relation instance $r_{2.6}$ satisfies fhd_1, since $r_{2.6} = \pi_{XY_1}(r_{2.6}) \bowtie \pi_{XY_2}(r_{2.6}) \bowtie \pi_{X(R-XY_1Y_2)}(r_{2.6})$, where $X =$

Table 2.6 Relation instance $r_{2.6}$ of DVD

	Title	Actor	Feature	Language
t_1	King Kong	Naomi Watts	Deleted Scenes	English
t_2	King Kong	Jack Black	Photo Gallery	English
t_3	King Kong	Naomi Watts	Photo Gallery	English
t_4	King Kong	Jack Black	Deleted Scenes	English
t_5	King Kong	Naomi Watts	Deleted Scenes	French
t_6	King Kong	Jack Black	Photo Gallery	French
t_7	King Kong	Naomi Watts	Photo Gallery	French
t_8	King Kong	Jack Black	Deleted Scenes	French

Table 2.7 A decomposed relation instance of DVD in $r_{2.6}$ (Title $\twoheadrightarrow$ Actor)

	Title	Actor
t_1	King Kong	Naomi Watts
t_2	King Kong	Jack Black

Table 2.8 A decomposed relation instance of DVD in $r_{2.6}$ (Title $\twoheadrightarrow$ Feature)

	Title	Feature
t_1	King Kong	Deleted Scenes
t_2	King Kong	Photo Gallery

Table 2.9 A decomposed relation instance of DVD in $r_{2.6}$ (Title $\twoheadrightarrow$ Language)

	Title	Language
t_1	King Kong	Englist
t_2	King Kong	French

{Title}, $Y_1 = $ {Actor}, $Y_2 = $ {Feature}, $R - XY_1Y_2 = $ {Language}. We can note that, the title of a DVD determines the set of actors independently from the rest of the information in any schema, and the title also determines the set of DVD features independently from the rest of the information in any schema. That is, the FHD decomposes $r_{2.6}$ into multiple new relations without loss of information.

Special Case: MVDs

As aforesaid, we can see from Fig. 1.1 that MVDs are special cases of FHDs. All MVDs can be represented as special FHDs with only one set of dependent attributes Y. Take mvd_1 in Sect. 2.9 as an example again. It can be rewritten as

$$fhd_2 : address, rate : \{region\}.$$

The relation instance $r_{2.3}$ in Table 2.3 satisfies the FHD, because $r_{2.3} = \pi_{XY}(r_{2.3}) \bowtie \pi_{X(R-XY)}(r_{2.3})$, where $X = $ {address, rate}, $Y = $ {region}. That is, the FHD decomposes $r_{2.3}$ into two new relations in Tables 2.4 and 2.5 without loss of information. In this example, FHD is regarded as an MVD. That is, FHDs subsume the semantics of MVDs, or FHDs generalize/extend MVDs, denoted by the arrow from MVDs to FHDs in Fig. 1.1.

Application

As mentioned in [48], FHDs can be used in the conceptual design of relational database logical schema. The properties of new type of dependencies relationship and links with the normalization process of relationship are studied. Through the normalization process, the relationship between the concept of first-order hierarchical decomposition and the concept of data hierarchical organization is discussed. FHDs are also studied in the presence of null values [92]. To explore all possible extensions from an incomplete database to a complete database, a possible world semantics can be used.

2.11 Approximate Multivalued Dependencies (AMVDs)

Similar to approximate functional dependencies (AFDs), approximate multivalued dependencies (AMVDs) [106] capture MVDs that approximately or almost hold in a relation. MVDs are very sensitive to noise, since a single incorrect or missing tuple may lead to the invalid schema. AMVDs are defined as ε-MVDs with the accuracy threshold $\varepsilon \geq 0$. The accuracy relates to the percentage of spurious tuples that will be introduced by joining the relations decomposed referring to the MVDs.

Definition

An *approximate multivalued dependency* (AMVD) is defined as ε-MVDs,

$$\text{AMVD} : X \twoheadrightarrow_\varepsilon Y,$$

where $X \cup Y \cup Z = R$, $\mathcal{J}(X \twoheadrightarrow Y_1) \leq \varepsilon$, and $\mathcal{J}$ is an entropic measure [14] of the schema with $\mathcal{J}(X \rightarrow Y|Z) = I(Y; Z|X)$. It follows $I(Y; Z|X) = H(XY) + H(XZ) - H(XYZ) - H(X)$, and $H(X) = \sum_{x \in \mathcal{D}} p(x) \log \frac{1}{p(x)}$, $H(X_1 X_2) = \sum_{x_1 \in \mathcal{D}_1, x_2 \in \mathcal{D}_2} p(x_1, x_2) \log \frac{1}{p(x_1, x_2)}$, where $p(x_1, x_2)$ is the joint distribution of two random variables (X_1, X_2), and $p(x_1), p(x_2)$ is the marginal distribution.

Example

For $r_{2.10}$ in Table 2.10, we can easily find that it does not satisfy the MVD child_name $\twoheadrightarrow$ phone, because the decomposition of Table 2.10 into Tables 2.11 and 2.12 referring to this MVD is not lossless, in other words, $r_{2.10} \neq \pi_{XY}(r_{2.10}) \bowtie \pi_{XZ}(r_{2.10})$, where $X = \{\text{ID}\}$, $Y = \{\text{phone}\}$, $Z = \{\text{child\_name}\}$. However, we can compute $\mathcal{J}(X \rightarrow Y|Z) = I(Y; Z|X) = 1/2$, since $H(X) = 1$, $H(XY) = 2$, $H(XZ) = 3/2$, $H(XYZ) = 2$. In this sense, given the threshold $\varepsilon = 0.5$, the relation in Table 2.10 can satisfy the AMVD

$$\text{amvd}_1 : \text{ID} \twoheadrightarrow_{0.5} \text{phone}.$$

To be specific, with tuple t_4 which has a different child_name of "William", the relation cannot satisfy the MVD above, but it is an AMVD with the approximate threshold $\varepsilon = 0.5$.

Table 2.10 Relation instance $r_{2.10}$ of AMVD

	Child_name	ID	Phone
t_1	David	99999	512-555-1234
t_2	David	88888	512-555-1234
t_3	David	99999	512-555-4321
t_4	William	88888	512-555-4321

Table 2.11 A decomposition {ID}{phone} for relation instance $r_{2.10}$ of AMVD

	ID	Phone
t_1	99999	512-555-1234
t_2	88888	512-555-1234
t_3	99999	512-555-4321
t_4	88888	512-555-4321

Table 2.12 A decomposition {ID}{child_name} for relation instance $r_{2.10}$ of AMVD

	Child_name	ID
t_1	David	99999
t_2	David	88888
t_3	William	88888

Special Case: MVDs

As aforesaid, when $\varepsilon = 0$, it is an exact MVD. Therefore, AMVD is a generalization of MVD with $\varepsilon \geq 0$. The mvd$_1$ in Sect. 2.9 can be also rewritten as

$$\text{amvd}_2 : \text{address, rate} \twoheadrightarrow_0 \text{region.}$$

We can find that this AMVD is an MVD with $\mathcal{J}(X \to Y|Z) = I(Y; Z|X) = 0$, since $H(X) = 3/2$, $H(XY) = 2$, $H(XZ) = 3/2$, $H(XYZ) = 2$, where $X = \{\text{address, rate}\}$, $Y = \{\text{region}\}$, $Z = \{\text{name}\}$. The threshold can be set as $\varepsilon = 0$ to express the exact MVD, and the entropic measure above satisfies the threshold. In this sense, the AMVDs subsume MVDs, in other words, AMVDs extend/generalize MVDs as denoted in Fig. 1.1 with arrow from MVDs to AMVDs.

Discovery

Kenig et al. [106] propose the algorithm MVDMiner for the discovery of approximate MVDs from data. To be specific, it mines for approximate MVDs with minimal separators. To discover the most specific sentences in the data that satisfy a certain criterion, such as the maximal items sets with the frequency in the data higher than the given threshold, the MVDMiner algorithm builds on previous results by Gunopulos et al. [90].

Application

As indicated in [106], AMVDs can be widely used in database and machine learning. For example, they can be applied in improvement of design, storage efficiency, performance of multiple aggregate queries and machine learning algorithms. With strict MVDs which are sensitive to noises, little errors can lead to invalidated schema. However, AMVDs have tolerance for these noises, which can keep more original schemas while detecting large invalidation.

2.12 Inclusion Dependencies (INDs)

Although integrity constraints such as FDs successfully capture data semantics, some data in the database may not meet such constraints. This is because data come from various independent sources (such as data integration), or involve complex, long-running activities, such as workflow. Instead of traditional FDs, inclusion dependencies [56, 57] use different relations to check the correctness of data.

Definition

An *inclusion dependency* (IND) can be defined as follows:

$$\text{IND} : R_1(X) \subseteq R_2(Y),$$

where X and Y are sets of attributes of relations R_1 and R_2, respectively. It means that for each tuple with attributes X in relation R_1, there must have another tuple with attributes Y in relation R_2.

Example

Take the instances in Tables 2.13 and 2.14, we can use the standard IND as follows:

$$\text{ind}_1 : \text{bookOrder(title, price)} \subseteq \text{book(title, price)}.$$

It means that for each tuple in the bookOrder relation, there must exist another tuple in relation book with same title and price. For example, in Table 2.14, tuple t_2 has title 'Harry

Table 2.13 Relation instance $r_{2.13}$ of book

	Title	Price
t_1	Harry Potter	17.99
t_2	Snow White	7.99

Table 2.14 Relation instance $r_{2.14}$ of bookOrder

	Title	Price
t_1	Snow White	7.99
t_2	Harry Potter	17.99

Potter' and price '17.99'. Then, there must exist another tuple t_3 in Table 2.13, that has the same title and price.

Axiomatization

As stated in [32], the axiomatization of INDs consists of the following rules:

IND1 (Reflexivity): $R[X] \subseteq R[X]$, if X is a sequence of distinct attributes of R.

IND2 (Projection and Permutation): if $R[A_1, ..., A_m] \subseteq S[B_1, ..., B_m]$, then $R[A_{i_1}, ..., A_{i_k}] \subseteq S[B_{i_1}, ..., B_{i_k}]$, for each sequence $i_1, ..., i_k$ of distinct integers from $\{1, ..., m\}$.

IND3 (Transitivity): if $R[X] \subseteq S[Y]$ and $S[Y] \subseteq T[Z]$, then $R[X] \subseteq T[Z]$.

The inference rules of INDs are complete and sound as proved in [32]. IND1 and IND3 generalize the Armstrong's axioms. IND2 says that any projections of permutations still satisfy the original INDs.

Discovery

As studied in [52], there are several algorithms for INDs discovery. In general, these discoveries can be divided into categories of unary and n-ary discoveries.

For unary IND discovery, the main idea of pruning is to use logical reasoning on INDs that have been discovered, along with basic column statistics, to avoid many expensive candidate checks. Some other algorithms with techniques, such as inverted index, disk-based sort-merge-join, deriving attributes clusters and using different partitioning schemes, can improve the efficiency of discovery.

The n-ary IND algorithms start with the unary INDs. The idea is to traverse the search space with candidate lattice as the model. To generate n-ary IND candidates, algorithm MIND proposes an approach based on apriori-gen. It uses the anti-monotony property of candidates to traverse the lattice bottom-up, and uses their invalid generalization (upward-pruning) to prune invalid candidates from the lattice. The extensions with some other methods, such as downward pruning, bottom-up and top-down traversals, modeling candidates as a hypergraph, unary IND coordinates, SQL for candidate validation, Apriori candidate generation, show a further improvement of discoveries.

Application

As mentioned in Chomicki and Marcinkowski [37], INDs can be used in data inconsistency detection and integrity restoration. When inconsistency is detected, one may try to keep as many tuples as possible, and recovery should have minimal impact on the database. This scenario is now called "minimum change integrity maintenance". Depending on whether the information in the database is considered correct and complete, one can interpret the minimum change hypothesis in several different ways. When the information in the database is complete but not necessarily correct, the database repair is thus to delete parts of the database. When the information in the database is incorrect and incomplete, one may consider both insertion and deletion.

2.13   Approximate Inclusion Dependencies (AINDs)

The interest of discovering strictly satisfied INDs may be limited in practice. Similar to the statistical extensions of FDs such as SFDs, PFDs or AFDs, the idea is to define an error measure from which approximate INDs can be rigorously defined with respect to a user-defined threshold. With the approximate INDs [124, 130], more violations and errors can be found.

Definition

An *approximate inclusion dependency* (AIND) can be defined as follows

$$\text{AIND} : R_i(X) \subseteq_\epsilon R_j(Y),$$

where the threshold ϵ is for an error measure called g_3', i.e., $g_3'(R_i(X) \subseteq R_j(Y)) \le \epsilon$. To be specific, with d for a database over a database schema $\mathbf{R}$ and $r_i, r_j \in d$ for relations over $R_i, R_j \in \mathbf{R}$ respectively, the following g_3' calculates the ratio of the minimum number of tuples that need to be removed from r to make $R_i(X) \subseteq R_j(Y)$ hold on r,

$$g_3'(R_i(X) \subseteq R_j(Y)) = 1 - \frac{\max\{|\pi_X(r')| \mid r' \subseteq r_i, (d - \{r_i\}) \cup \{r'\} \models R_i(X) \subseteq R_j(Y)\}}{|\pi_X(r_i)|}.$$

Here r' is a subset of tuples in r_i that do not violate $R_i(X) \subseteq R_j(Y)$, denoted by $(d - \{r_i\}) \cup \{r'\} \models R_i(X) \subseteq R_j(Y)$. The measure g_3' defines the approximation of an inclusion dependency $R_i(X) \subseteq R_j(Y)$. A natural interpretation is the fraction of rows with exceptions or errors affecting the dependencies. The smaller the g_3' measure is, the more likely it becomes an IND.

Example

For an example in Tables 2.15 and 2.16, we have

$$g_3'(R_1(\text{address1}) \subseteq R_2(\text{address2})) = 1 - 1/2 = 0.5,$$

$$g_3'(R_2(\text{address2}) \subseteq R_1(\text{address1})) = 1 - 1/3 = 0.66.$$

Referring to the previous definition, a better inclusion dependency is from $R_1(\text{address1})$ to $R_2(\text{address2})$ with a lower error measure. When we set $\epsilon = 0.5$, the AIND can be obtained as

$$\text{aind}_1 : R_1(\text{address1}) \subseteq_{0.5} R_2(\text{address2}).$$

Table 2.15 Relation instance $r_{2.15}$ of R_1

	Address1	New address1
t_1	Central Park	Central Park
t_2	Central Park	West Lake
t_3	Central Park	Fifth Avenue
t_4	West Lake	New Center

Table 2.16 Relation instance $r_{2.16}$ of R_2

	Address2
t_5	West Lake
t_6	Fifth Avenue
t_7	New Center

It states that $R_1(\text{address1}) \subseteq R_2(\text{address2})$ is more likely to hold. By removing 3 tuples, the IND holds.

Special Case: INDs

As mentioned in Fig. 1.1, AINDs extend INDs with the threshold ϵ for g_3'. When we set $\epsilon = 0$, i.e., if there is an AIND with $g_3'(R_1(X) \subseteq R_2(Y)) = 0$, then $R_1(X) \subseteq R_2(Y)$ is an exactly IND. For example, the ind_1 in Sect. 2.12 can be rewritten as an AIND as follows:

$$\text{aind}_2 : \text{bookOrder(title, price)} \subseteq_0 \text{book(title, price)}.$$

Therefore, AINDs subsume the INDs, or AINDs generalize/extend INDs, denoted by the arrow from INDs to AINDs in Fig. 1.1.

Discovery

In the setting of approximate IND discovery, Lopes et al. [124] propose to consider some data inconsistencies. The NavLog' discovery algorithm has two steps in general to cope with approximate inclusion dependencies. First, it determines the best direction between two attribute sequences, computed from projections. Second, the error g_3' is computed from its definition.

Application

As mentioned in [124], they propose a novel application, *self-tuning the logical database design*, to use the discovered approximate inclusion dependencies effectively. We can use AINDs to detect missing constraints, data integrity problems or misconception of a database. AINDs are considered with "almost hold" in a database. They can be employed to reduce the tasks of database administration. Based on these ideas, a prototype DBA companion has been implemented. As mentioned in [130], a new challenge for the database community about simplifying database administration has been recognized.

2.14 Conditional Inclusion Dependencies (CINDs)

Analogous to CFDs extending FDs, Bravo et al. [26], Ma et al. [126] study conditional inclusion dependencies (CINDs), which are inclusion dependencies that hold only on a subset of tuples. That is, CINDs extend traditional inclusion dependencies (INDs) by enforcing bindings of semantically related data values. They are useful not only in data cleaning but also in contextual schema matching.

Definition

A *conditional inclusion dependency* (CIND) is a pair

$$\text{CIND} : R_1[X; X_p] \subseteq R_2[Y; Y_p], t_p,$$

where (1) X, X_p and Y, Y_p are sets of attributes of relations R_1 and R_2, respectively, such that X and X_p (resp. Y and Y_p) are disjoint, and there can be a *nil* to express empty; (2) $R_1(X) \subseteq R_2(Y)$ is a standard IND, referred to as the IND embedded in CIND; (3) t_p is a pattern tuple with attributes in X, X_p and Y, Y_p, where for each $A \in X \cup X_p \cup Y \cup Y_p$, $t_p[A]$ is either a constant 'a' in $dom(A)$, or an unnamed variable '_' that draws values from $dom(A)$.

Example

For example, a CIND in Tables 2.13 and 2.17 could be

cind$_1$: order[title, price, type; *nil*] $\subseteq$ book[title, price; *nil*], (_, _, 'book'||_, _),

which can also be written in another format

$$\text{order(title, price, type = 'book')} \subseteq \text{book(title, price)}.$$

It states that for each tuple t in the order relation, if the type of t is book, then there must exist a tuple t' in book such that $t[\text{title}] = t'[\text{title}]$ and $t[\text{price}] = t'[\text{price}]$. In other words,

Table 2.17 Relation instance $r_{2.17}$ of order

	Title	Type	Price
t_1	Pride and Prejudice	CD	21.99
t_2	Harry Potter	Book	17.99

instead of the entire relation, this constraint is an inclusion dependency that holds only on the subset tuples whose type = 'book'. As shown in Table 2.17, the type of t_2 is 'book', the title is 'Harry Potter' and the price is '17.99'. Then, in Table 2.13, there must be a tuple that has the same title and price, i.e., t_3.

Special Case: INDs

As mentioned in Fig. 1.1, CINDs extend INDs. In other words, CINDs are special cases of INDs. The ind_1 in Sect. 2.12 can be represented to a CIND with conditional pattern tuple as follows:

$$cind_2 : bookOrder[title, price; nil] \subseteq book[title, price; nil], (\_, \_||\_, \_).$$

It means that for each tuple t with any title and price in relation order, there must exist a tuple t' in book such that $t[title] = t'[title]$ and $t[price] = t'[price]$. Consequently, CINDs subsume INDs, or CINDs generalize/extend INDs, denoted by the arrow from INDs to CINDs in Fig. 1.1.

Discovery

Curé [45] proposes an algorithm for CINDs discovery, which starts from a set of approximated INDs and finds pattern tuples to turn them into CINDs. Bauckmann et al. [11] adopt similar strategies and present two algorithms. CindeRELLA employs a breadth-first traversal, while PLI, leveraging value position lists, employs a depth-first traversal, based on an Apriori algorithm. Kruse et al. [113] propose a distributed system RDFind to solve the CINDs discovery problem over RDF datasets. RDFind can discover all CINDs that are both minimal and broad in a given RDF dataset. To improve efficiency, it uses a lazy pruning strategy to reduce the search space.

Application

Bravo et al. [26] propose heuristic algorithms for the consistency analysis of CINDs and show that CINDs are useful in contextual schema matching. Each match is annotated with a logical condition providing the context in which the match should apply. Fan et al. [64] give methods for data repairing and record matching using CINDs. They develop a prototype system Semandaq for improving the quality of relational data. To be specific, based on CINDs, it can discover data quality rules, detect errors and inconsistencies. Then, CINDs support data repairing and record matching.

2.15   Summary and Discussion

The extensions of conventional dependencies over categorical data are in four subcategories, i.e., statistical, conditional, multivalued and inclusion. In this section, we summarize and compare the extension relationships of data dependencies, in four subcategories. Note that the

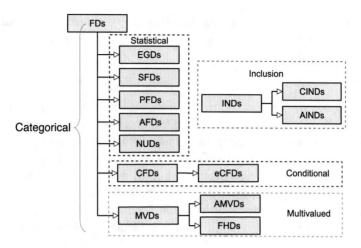

Fig. 2.2 A family tree of extensions between categorical data dependencies

extension relationships of different dependencies were introduced and described separately in the corresponding sections. In order to obtain a global picture of all these extension relationships, Fig. 2.2 plots them in four subcategories. The comparison of the extension relationships is thus discussed in each subcategory. It is worth noting that most of these extensions are still based on the equality relationship of data values, i.e., not effective in addressing the various information formats on heterogeneous data.

Statistical

The extensions of SFDs [98] in Sect. 2.3, PFDs [167] in Sect. 2.4, AFDs [109] in Sect. 2.5, and NUDs [89] in Sect. 2.6 are with statistics. Instead of the FDs exactly holding in the data, it is to find almost valid FDs. That is, these FD extensions are applicable to the workload where FDs hold in most tuples in a relation. Given more (approximate) rules, the recall of violation detection can be improved, while it may drag down the precision. In particular, SFDs are efficient to compute by domain, while AFDs can tell a fine-grain proportion of violation tuples. That is, AFDs could more accurately specify the proportion of violation tuples and get the minimum number of tuples that need to be removed to satisfy standard FDs. Similar to SFDs, NUDs count the maximum number of Y values associated to each X value. Furthermore, PFDs study normalization with domain instead of tuples.

Conditional

Different from the aforesaid statistical extensions that are still declared over the whole relation, the extensions with conditions, i.e., CFDs [22] and eCFDs [25] in Sects. 2.7 and 2.8, can be used in a workload that FDs hold only in a part of the relation. While CFDs extend FDs with conditions, to identify the subgroup of tuples where FDs hold, eCFDs further extend CFDs with $\neq, <, \leq, >$ and $\geq$ predicates instead of only equality $=$. Unlike the approximately holding FDs, the accurately declared CFDs naturally have a high precision of violation detection. The coverage (recall), however, is limited, since the data dependencies hold only in a part of the data.

Multivalued

While FDs rule out the existence of certain tuples, known as equality-generating dependencies, MVDs [13, 55] require the presence of certain tuples, i.e., tuple-generating dependencies. Rather than X exactly determines Y, the multivalued dependencies state that X multidetermines Y, i.e., a particular X value is associated with a set of Y values and $R - YZ$ values, and these two sets are independent of each other. While MVDs decompose a relation into two of its projections without loss of information, FHDs [48, 92] further study the hierarchical decomposition of a relation into multiple relations. Similar to AFDs extending FDs as aforesaid, AMVDs [106] are MVDs that approximately hold in a relation.

Inclusion

Finally, INDs [56, 57] extend the constraints between two relations, i.e., the constraints on existence. Again, the interest of discovering strictly satisfied INDs may be limited in practice. Similar to the statistical extensions of FDs such as SFDs, PFDs or AFDs, the idea of AINDs [124, 130] is to define an error measure from which approximate INDs can be rigorously defined with respect to a user-defined threshold. Likewise, analogous to CFDs extending FDs, CINDs [26, 126] are INDs that hold only on the subset of tuples. CINDs extend traditional INDs by enforcing bindings of semantically related data values.

Heterogeneous Data

Extensions with statistics or conditions to FDs are still based on equality constraints. Despite the enhanced definitions, it is not robust enough to address the variety issues of big data. Data obtained from merging heterogeneous sources often have various representation conventions. Dataspaces [81, 91] provide a co-existing system of heterogeneous data from multiple sources. Carefully declared data dependencies over the heterogeneous data would be useful in query optimization and consistent query answering in dataspaces.

Table 3.2 presents some example data in different formats, where the tuples are from two heterogeneous sources s_1 and s_2. For instance, "12th St." in tuple t_5 from source s_2 and "12th Str" in tuple t_6 from source s_1 denote the same street but with different formats.

To capture the slight difference between values, the distance functions are applied to evaluate how similar the two values are. For numeric attributes, absolute value of difference $d(x_1, x_2) = |x_1 - x_2|$ is a common distance metric. For string/text attributes, Elmagarmid et al. [54] provide a survey about similarity metrics, such as editing distance (character-based), q-gram (token-based), soundex (phonetic-based), etc.

The heterogeneous part in Fig. 1.1 presents the major data dependencies defined over heterogeneous data. The corresponding special cases, where the data dependencies are extended from, are presented as well. Each pair of a data dependency and its special case, e.g., MFDs extend FDs, corresponds to an arrow from FDs to MFDs in the family tree in Fig. 1.1.

Table 3.1 further categorizes the extensions in the following aspects. (1) Metric: To tolerate slight differences in heterogeneous data, a natural extension is to introduce a distance/similarity metric, in either the determinant or dependent attributes (or both) of the data dependencies. (2) Additional: To address more complicated scenarios, additional extensions are introduced to the distance metric, e.g., fuzzy resemblance relation is employed to replace the FDs' equality comparison on domain values with "approximately equal", "more or less equal", etc. (3) Identification: Finally, the data dependencies are used as matching rules, i.e., given certain X values, it is sufficient to imply the identification on Y.

© The Author(s), under exclusive license to Springer Nature Switzerland AG 2023   47
S. Song and L. Chen, *Integrity Constraints on Rich Data Types*,
Synthesis Lectures on Data Management,
https://doi.org/10.1007/978-3-031-27177-9_3

Table 3.1 Category and special cases of data dependencies over heterogeneous data, where each pair corresponds to an arrow in Fig. 1.1 such as CMDs◁–MDs

Subcategory	Sections	Data dependencies	Special cases
Metric	3.1	MFDs	FDs
Metric	3.2	NEDs	MFDs
Metric	3.3	DDs	NEDs
Additional-condition	3.4	CDDs	DDs, CFDs
Additional-attribute	3.5	CDs	NEDs
Additional-probability	3.6	PACs	NEDs
Additional-fuzzy	3.7	FFDs	FDs
Identification	3.8	ONFDs	FDs
Identification	3.9	MDs	FDs
Identification	3.10	CMDs	MDs

3.1   Metric Functional Dependencies (MFDs)

In the case of different data sources, small changes in data format and interpretation without inherent semantic conflicts will violate traditional FDs. Therefore, it is necessary to generalize the traditional FDs by using metric function dependence, which allows slight differences in the values of FD dependent attributes Y (controlled by metrics). Metric functional dependencies (MFDs) [112] extend FDs with distance/similarity metrics on the dependent attributes Y, given the exactly matched (i.e., equal) values on determinant attributes X.

Definition

A *metric functional dependency* (MFD) over R has the form

$$\text{MFD} : X \rightarrow {}^{\delta}Y,$$

where (1) X, Y are two sets of attributes in R; (2) $\delta \geq 0$ is a threshold of metric distance on attributes in Y. The metric d is defined on the domain of Y, i.e., $d : dom(Y) \times dom(Y) \rightarrow \mathbb{R}$. A relation instance r over schema R satisfies the MFD, if any two tuples $t_1, t_2 \in r$ having $t_1[X] = t_2[X]$ must have distance $\leq \delta$ on attributes in Y.

Example

Consider an example relation instance in Table 3.2.
    An MFD over the relation $r_{3.2}$ can be

$$\text{mfd}_1 : \text{name, region} \rightarrow^{500} \text{price.}$$

Table 3.2 Relation instance $r_{3.2}$ with tuples from heterogeneous sources

	Source	Name	Street	Address	Region	Zip	Price	Tax
t_1	s_1	NC	Central Park	#5, Central Park	New York	10041	299	29
t_2	s_2	NC	12th St.	#2 Ave, 12th St.	San Jose	95102	300	20
t_3	s_1	Regis	Central Park	#9, Central Park	New York	10041	319	31
t_4	s_2	Chris	61st St.	#5 Ave, 61st St.	Chicago	60601	499	49
t_5	s_2	WD	12th St.	#6 Ave, 12th St.	San Jose	95102	399	27
t_6	s_1	NC	12th Str	#2 Aven, 12th St.	San Jose	95102	300	20

It states that if two tuples such as t_2 and t_6 have identical name and region, then their distance on attribute price should be ≤ 500, i.e., close rather than exactly equal.

Special Case: FDs

As indicated in Fig. 1.1, MFDs extend FDs. When $\delta = 0$, an MFD states that if two tuples have equal X values, then their A values have distance 0, i.e., equal as well. It is exactly an FD. In other words, all FDs can be represented as special MFDs with $\delta = 0$. For example, fd_1 in Sect. 1.1 can be equivalently represented by

$$\mathsf{mfd}_2 : \mathsf{address} \to^0 \mathsf{region}.$$

That is, for any two tuples, equal address implies region value distance 0. Consequently, MFDs subsume the semantics of FDs, or MFDs generalize/extend FDs, denoted by the arrow from FDs to MFDs in Fig. 1.1.

Discovery

During the MFDs discovery, a key step is to verify whether a candidate MFD holds for a given relation [112]. Similar to the computation of g_3 measure for AFDs in Sect. 2.5, it first groups all the tuples according to the LHS attributes X. For each group of tuples with equal X values, the maximum distance between any two tuples is computed, known as the diameter. Obviously, the MFD holds if the diameter has $\leq \delta$ in each group. In $O(n^2)$ time one can verify whether an MFD holds, where $n = |r|$ is the size of the relation r.

Koudas et al. [112] also give a linear time approximate verification algorithm that verifies whether an MFD holds for a given instance. Given an MFD $X \to^\delta A$ and an error parameter $\epsilon \geq 0$, the algorithm first computes an approximate diameter (i.e., the maximum distance) on attribute A within a $(1 + \epsilon)$ factor for each partition of the instance with respect to X.

When an MFD $X \rightarrow {}^{\delta}A$ holds in the instance, it can be verified that the MFD $X \rightarrow {}^{\delta'}A$ does not hold in the instance , where $\delta' > \delta(1 + \epsilon)$.

Application

MFDs are useful in violation detection. For instance, as presented in [112], one might expect a functional dependency of the form

$$\text{address} \rightarrow (\text{latitude, longitude})$$

to hold. It is worth noting that the variations are a natural part of the location data and cannot be eliminated by format standardization. With MFDs, such small variations will not be detected erroneously as violations. The MFDs violation detection, where similarity metrics are considered, cannot adopt the efficient operators, such as GROUP BY identification data in SQL. According to the observation, given a general metric space of dependent (right-hand-side) attributes Y, it takes $O(n^2)$ time to detect violations, where n is the total number of tuples. Therefore, approximation approaches are developed by relaxing the similarity thresholds in Y.

As described in [137], MFD can be used for data quality problems, they can identify inconsistencies in the domain by measuring the huge differences in the data and correct the inconsistent data to minimize the repair of static distortion. In order to identify semantically related attributes, they propose a perfect axiomatic method and an effective algorithm for MFDs implication testing. Combining the quantitative method and logical method, this paper introduces a new constraint-based cleaning strategy, which uses statistical distortion to ensure the highest quality of the selected (minimum) repair.

3.2   Neighborhood Dependencies (NEDs)

Neighborhood dependencies (NEDs) [10] declare data dependencies on the closeness of neighbor attribute values. While MFDs introduce distance metric in the dependent attributes Y, NEDs consider metrics on both sides of attributes. Intuitively, it states that if two tuples are close with respect to the predictor variables (determinant attributes), then the two tuples should have similar values for the target variable (dependent attribute).

Definition

A *closeness function* $\theta_A(\cdot, \cdot)$ is associated to each attribute $A \in R$. The inputs of $\theta_A(\cdot, \cdot)$ are two values of attribute A, and the output is a number denoting the distance/similarity of the two input values.

A *neighborhood predicate* specifies thresholds of distance/similarity (the original definition is similarity, and for convenience, we use distance by default) on attributes,

$$A_1^{\alpha_1} \dots A_n^{\alpha_n},$$

where (1) A_i, $1 \leq i \leq n$ are attributes of a relation R; (2) $\alpha_i \geq 0$, $1 \leq i \leq n$ are thresholds of distance/similarity on corresponding attributes A_i. Two tuples t_1, t_2 agree on the predicate if their similarity on each attribute A_i satisfies $\theta_{A_i}(t_1[A_i], t_2[A_i]) \geq \alpha_i$, or their distance on each attribute A_i satisfies $\theta_{A_i}(t_1[A_i], t_2[A_i]) \leq \alpha_i$.

A *neighborhood dependency* (NED) declares constraints between two neighborhood predicates

$$\text{NED} : A_1^{\alpha_1} \ldots A_n^{\alpha_n} \rightarrow B_1^{\beta_1} \ldots B_m^{\beta_m}.$$

A relation instance r satisfies the constraints, if for each pair of tuples in r that agrees on the predicate $A_1^{\alpha_1} \ldots A_n^{\alpha_n}$, they satisfy the predicate $B_1^{\beta_1} \ldots B_m^{\beta_m}$ as well.

Example

For tuples in Table 3.2, a neighborhood predicate can be

$$\text{name}^1 \text{address}^5.$$

It specifies distance thresholds 1 and 5 on attributes name and address, respectively. Two tuples t_2 and t_6 are said agreeing on the predicate, since their edit distances [132] have $\theta_{\text{name}}(t_2[\text{name}], t_6[\text{name}]) = 0 \leq 1$ and $\theta_{\text{address}}(t_2[\text{address}], t_6[\text{address}]) = 1 \leq 5$.

An NED with distance thresholds is declared by

$$\text{ned}_1 : \text{name}^1 \text{address}^5 \rightarrow \text{street}^5.$$

It states that two tuples such as t_2 and t_6 mentioned above, having similar names and addresses, should have similar streets as well, i.e., $\theta_{\text{street}}(t_2[\text{street}], t_6[\text{street}]) = 1 \leq 5$.

Special Case: MFDs

In Fig. 1.1, we can find that NEDs extend MFDs. When the distance threshold has $\alpha_i = 0$ in an NED, it denotes equality constraints on attributes X. Consider the example mfd$_1$ in Sect. 3.1, it can be represented by an NED as follows:

$$\text{ned}_2 : \text{name}^0 \text{region}^0 \rightarrow \text{price}^{500}.$$

It states that if two tuples have the same name and region, then their distance of price should be ≤ 500. In other words, all MFDs can be represented as special NEDs with distance thresholds $\alpha_i = 0$. In this sense, NEDs subsume the semantics of FDs, or NEDs generalize/extend FDs, denoted by the arrow from FDs to NEDs in Fig. 1.1.

Discovery

The NEDs discovery problem [10] is given the target right-hand side neighborhood predicate, to find a left-hand side neighborhood predicate such that the resulting NED has sufficient support and confidence. It is proved to be NP-hard in terms of the number of attributes.

Application

NEDs can be used for filling unknown values or repairing data errors, by a P-neighborhood method [10]. Let P and T denote the left-hand-side predicator attributes and the right-hand side target attributes of an NED, respectively. The P-neighborhood method predicts the T value of a new tuple based on all existing neighbors of the tuple under the closeness on P attributes. This proposal is more intuitive, since the k-nearest-neighbor (kNN) method does not predefine an appropriate distance metric or k.

3.3   Differential Dependencies (DDs)

In addition to the "similar" semantics considered by MFDs in Sect. 3.1 and NEDs in Sect. 3.2, "dissimilar" semantics could also be useful for declaring dependencies between different types of data, such as numeric or textual values. For instance, if two transactions of a card are recorded in distant cities, then their transaction time should also be distant, e.g., in different days. In this sense, Song and Chen [146] propose differential dependencies (DDs), introducing some advanced differential functions to express the constraints on distances. That is, DDs consider constraints not only on similarity but also on dissimilarity. In short, if two tuples have distances on attributes X agreeing with a certain differential function, then their distances on attribute A should also agree with another differential function.

Definition

A *similarity/distance metric*, d_A, is associated to each attribute A, which satisfies non-negativity, identity of indiscernible and symmetry. For example, the metric on a numerical attribute can be the absolute value of difference, i.e., $d_A(a, b) = |a - b|$. For a text attribute, we can adopt string similarity such as edit distance (see [132] for a survey).

A *differential function* $\phi[A]$ on attribute A indicates a range of metric distances, specified by $\{=, <, >, \leq, \geq\}$. Two tuples t_1, t_2 are said compatible w.r.t. differential function $\phi[A]$, denoted by $(t_1, t_2) \asymp \phi[A]$, if the metric distance of t_1 and t_2 on attribute A is within the range specified by $\phi[A]$, i.e., satisfying/agreeing with the distance restriction $\phi[A]$. A differential function may also be specified on a set of attributes X, denoted by $\phi[X]$. It denotes a pattern of differential functions (distance ranges) on all the attributes in X.

A *differential dependency* (DD) over a relation R is a form of

$$\text{DD} : \phi[X] \rightarrow \phi[Y],$$

where $X, Y \subseteq R$ are attributes in R. The dependency states that any two tuples satisfying differential function $\phi[X]$ must satisfy $\phi[Y]$.

A DD is in *standard form*, if $\phi[Y]$ is specified on a single dependent attribute Y. According to the decomposition rule in the inference system, a non-standard form DD can be equivalently split into multiple DDs in standard form [146].

We say that a relation I of $\mathcal{R}$ *satisfies* a DD, denoted by $I \vDash (X \rightarrow A, \phi[XA])$, if for any two tuples t_1 and t_2 in I, $(t_1, t_2) \asymp \phi[X]$ implies $(t_1, t_2) \asymp \phi[A]$. Likewise, a relation I satisfies a set Σ of DDs, $I \vDash \Sigma$, if I satisfies each DD in Σ.

Example

Consider a differential dependency in Table 3.2,

$$dd_1 : \mathsf{name}(\leq 1), \mathsf{street}(\leq 5) \rightarrow \mathsf{address}(\leq 5).$$

It states that if two tuples such as t_2 and t_6 have similar names (i.e., having edit distance 0 on name in the range of $[0, 1]$) and street values (with distance 1 in $[0, 5]$), they must share similar addresses values as well (having address value distance 1 in $[0, 5]$).

To express the semantics on "dissimilar", a differential dependency could be

$$dd_2 : \mathsf{street}(\geq 10) \rightarrow \mathsf{address}(\geq 5).$$

That is, for any two tuples, e.g., t_1 and t_2, whose distance on street is 10 (≥ 10), their distance on address must be greater than 5. In other words, if the streets of two tuples are not similar, the corresponding addresses should be dissimilar.

Special Case: NEDs

Figure 1.1 indicates that DDs extend NEDs. When the differential functions in a DD express only the "similar" semantics, it is exactly an NED. The example ned_1 in Sect. 3.2 can be represented by a DD as follows,

$$dd_3 : \mathsf{name}(\leq 1), \mathsf{address}(\leq 5) \rightarrow \mathsf{street}(\leq 5).$$

Thereby, DDs subsume the semantics of NEDs, or DDs generalize/extend NEDs, denoted by the arrow from NEDs to DDs in Fig. 1.1.

Axiomatization

For the class of heterogeneous data, Song and Chen [146] introduce a complete and sound set of inference rules for DDs. Let Σ be a set of DDs over relation R. For DDs, there are four inference rules as follows:

DD1 : If $Y \subseteq X$ and $\phi_L[Y] = \phi_R[Y]$, then $\Sigma \vdash_{\mathcal{I}} \phi_L[X] \rightarrow \phi_R[Y]$.

DD2 : If $\Sigma \vdash_{\mathcal{I}} \phi_L[X] \rightarrow \phi_R[Y]$, then $\Sigma \vdash_{\mathcal{I}} \phi_L[X] \wedge \phi_1[Z] \rightarrow \phi_R[Y] \wedge \phi_1[Z]$.

DD3 : If $\Sigma \vdash_{\mathcal{I}} \phi_L[X] \rightarrow \phi_1[Z]$, $\phi_1[Z] \leq \phi_2[Z]$ and $\Sigma \vdash_{\mathcal{I}} \phi_2[Z] \rightarrow \phi_R[Y]$, then $\Sigma \vdash_{\mathcal{I}} \phi_L[X] \rightarrow \phi_R[Y]$.

DD4 : If $\Sigma \vdash_{\mathcal{I}} \phi_L[X] \wedge \phi_i[B] \rightarrow \phi_R[Y]$, $1 \leq i \leq k$, and $\left(\Sigma, \overline{\phi_1[B]} \wedge \cdots \wedge \overline{\phi_k[B]} \right)$ is inconsistent, then $\Sigma \vdash_{\mathcal{I}} \phi_L[X] \rightarrow \phi_R[Y]$.

DD1, DD2 and DD3 are modified according to the Reflexivity, Augmentation and Transitivity rule in Armstrong's axioms, respectively. DD4 is specific to differential functions.

Discovery

Song and Chen [146] first introduce the concept of minimal DDs and indicate that even the discovered minimal DDs could be exponentially large in size w.r.t. to the number of attributes. Several pruning strategies are then devised for DDs discovery. (1) *Negative Pruning* explores the subsumption order relation in the search space, rather than evaluating all possible differential functions. (2) *Positive Pruning* utilizes the property that certain DDs may imply others for any relation instance, and it can be used in a top-down search procedure. (3) *Hybrid Pruning* can be used to avoid the worst cases for negative and positive pruning, and combine the two pruning in the way that, if the positive pruning is not valid, then the negative one is applied.

Song et al. [148, 149] study the determination of distance thresholds of differential functions in a parameter-free style. Moreover, Kwashie et al. [115] propose a subspace-clustering-based approach to discover DDs. The implication problem for DDs, i.e., implying a DD from a set of DDs, is co-NP-complete [146] .

Application

As indicated in [146], DDs can be used to rewrite queries in semantic query optimization. In data partition, DDs can reduce the number of predicates and improve the efficiency. In duplicate detection, DDs can address more matching rules by introducing various differential functions on one attribute. Furthermore, DDs are used as integrity constraints to enrich the candidates for missing data imputation [156, 157], and repairing data errors [153, 155].

3.4 Conditional Differential Dependencies (CDDs)

Similar to the extension of FDs to conditional FDs (CFDs) [22], an extension of DDs to hold over subsets of relations, called *Conditional Differential Dependencies* (CDDs), is also proposed [116]. Likewise, the necessity of extending DDs is that it can reserve data subset conditionally to capture some potential knowledge and inconsistencies. While the conditions in CDDs are categorical values, the distance constraints can be declared over heterogeneous data. For instance, a CDD may state that in the region of "Chicago" (categorical value), two tuples (from heterogeneous sources) with similar name values (denoting the same hotel) should have similar address values. In this sense, CDDs extend both DDs over heterogeneous data and CFDs over categorical data.

Definition

To declare the conditions, a predicate on attribute $A \in R$ is introduced, i.e., $\psi[A] = \{A \text{ op } a\}$. The operator op is one of the relation symbols of $\{=, <, >, \leq, \geq, \in\}$. The item a is a single value in $dom(A)$, and a can be a '_'.

A *conditional differential dependency* (CDD) is in the form

$$CDD : \phi[X] \rightarrow \phi[Y], \Psi[XY],$$

where (1) $X, Y \subseteq R$ are attributes in R; (2) $\phi[X] \rightarrow \phi[Y]$ is the embedded DD in CDD; and (3) $\Psi[XY]$ is a pattern of conditions. For each attribute $A \in X \cup Y$, $\psi[A]$ in $\Psi[XY]$ is either a single value in $dom(A)$ or a variable '_'. It specifies that any two tuples with the condition $\Psi[XY]$ and satisfying the differential function $\phi[X]$ must satisfy $\phi[Y]$ as well.

Example

Consider a CDD with conditions $\Psi(XY) = (NC, \_ \parallel \_)$ over attributes name, street and address as follows,

$$cdd_1 : name(\leq 0), street(\leq 5) \rightarrow address(\leq 5), (NC, \_ \parallel \_).$$

It states that if two tuples, such as t_2 and t_6 in $r_{3.2}$ of Table 3.2, have the same name "NC" and similar streets, i.e., having street edit distance in the range of $[0, 5]$, they must share similar address values (having address value distance in $[0, 5]$).

Special Case: DDs

As shown in Fig. 1.1, CDDs naturally extend DDs in Sect. 3.3. When the distance function on X determines that of Y without conditions, it is exactly a DD. In other words, all DDs can be represented as special CDDs without conditions. For the example dd_1 in Sect. 3.3, it can be expressed by a CDD as follows,

$$cdd_2 : name(\leq 1), street(\leq 5) \rightarrow address(\leq 5), (\_, \_ \parallel \_).$$

In this sense, CDDs subsume the semantics of DDs, or CDDs generalize/extend DDs, denoted by the arrow from DDs to CDDs in Fig. 1.1.

Special Case: CFDs

Figure 1.1 also states that CDDs extend CFDs in Sect. 2.7. When X determines Y with special differential functions, i.e., the distances on the attributes of two tuples are 0, it is exactly a CFD. In other words, all CFDs can be represented as special CDDs with distances 0. For example, cfd_1 in Sect. 2.7 can be represented as a CDD,

$$cdd_3 : region(= 0), name(= 0) \rightarrow address(= 0), (Jackson, \_ \parallel \_).$$

In this sense, CDDs subsume the semantics of CFDs, or CDDs generalize/extend CFDs, denoted by the arrow from CFDs to CDDs in Fig. 1.1.

Discovery

Kwashie et al. [116] present an algorithm, MineCDD, which mines a minimal cover set of CDDs holding in the given relational instance. The complexity is $O(n^2)$ for n as the size of relation. It consists of four phases: (1) generating conditional statements $\Psi[XY]$; (2) forming LHS differential functions $\phi[X]$; (3) discovering RHS differential functions $\phi[Y]$; and (4) pruning the found CDDs set.

Given the aforesaid DDs and CFDs as special cases, it is not surprising that inference rules for pruning in DD discovery in Sect. 3.3 and CFD discovery in Sect. 2.7 naturally apply in CDD discovery.

While CDDs have more expressiveness than CFDs and DDs, their inference rules are applicable to the CDD discovery. For example, the well-known Armstrong's Axioms [8] form the bases of the finite axiomatizability of FDs, as introduced at the beginning of Chap. 2. Similar axioms have been derived for CFDs and DDs. A few of the inference rules are used in the pruning for CDDs.

Application

CDDs can be applied in data quality management and knowledge discovery [116]. First, given the more advanced expressive power, CDDs could detect more errors that cannot be captured by CFDs with only equality operators. Moreover, more interesting local rules (dependencies) are obtained in addition to global rules, which can thus be utilized in classification, e.g., classifying more accurately the iris plants.

3.5 Comparable Dependencies (CDs)

Recall that Dataspaces [81, 91] provide a co-existing system of heterogeneous data from multiple sources. That is, heterogeneity exists not only on values but also on attribute names in the data collected from various sources in a dataspace.

Rather than a single relation, Song et al. [151, 152] define a general form of comparable dependencies (CDs), which specifies constraints on comparable attributes and various comparison operators of two relations from heterogeneous sources. CDs consider the matching of both heterogeneous attribute names and values.

Definition

A *similarity function*

$$\theta(A_i, A_j) : [A_i \approx_{ii} A_i, A_i \approx_{ij} A_j, A_j \approx_{jj} A_j]$$

specifies a constraint on similarity of two values from attribute A_i or A_j, according to the corresponding similarity operators $\approx_{ii}, \approx_{ij}$ or $\approx_{jj}$. Here, A_i, A_j are often synonym attributes, and the similarity function comes together with the attribute matching on how their attribute values should be compared. Two tuples t_1, t_2 are said to be similar w.r.t. $\theta(A_i, A_j)$, denoted by $(t_1, t_2) \asymp \theta(A_i, A_j)$, if at least one of three similarity operators in $\theta(A_i, A_j)$ evaluates to true.

A *comparable dependency* (CD) is in the form of

$$CD : \bigwedge \theta(A_i, A_j) \rightarrow \theta(B_i, B_j),$$

where $\theta(A_i, A_j)$ and $\theta(B_i, B_j)$ are similarity functions. It states that for any two tuples t_1, t_2 that are similar w.r.t. $\theta(A_i, A_j)$, it implies $(t_1, t_2) \asymp \theta(B_i, B_j)$ as well.

Example

For instance, consider a dataspace with 3 tuples,

$$t_1 : \{name : Alice, region : Petersburg, addr : \#7\ T\ Avenue\};$$
$$t_2 : \{name : Alice, city : St\ Petersburg, post : \#7\ T\ Avenue\};$$
$$t_3 : \{name : Alex, region : St\ Petersburg, post : No\ 7\ T\ Ave\}.$$

A similarity function specified on two attributes region and city can be

$$\theta(region, city) : [region \approx_{\leq 5} region, region \approx_{\leq 5} city,$$
$$city \approx_{\leq 5} city].$$

Two tuples agree the similarity function if either their region values have distance ≤ 5, or their city values have distance ≤ 5, or the region of a tuple is similar to the city of the other with distance ≤ 5. For instance, t_1 and t_2 have region and city value distance $2 \leq 5$, and thus agree $\theta(region, city)$.

Likewise, a similarity function on addr and post can be

$$\theta(addr, post) : [addr \approx_{\leq 7} addr, addr \approx_{\leq 9} post, post \approx_{\leq 5} post].$$

Tuples t_2 and t_3 having post values with distance $5 \leq 5$ again satisfy $\theta(region, city)$.

Consequently, a CD can be declared as

$$cd_1 : \theta(region, city) \rightarrow \theta(addr, post).$$

It states that if the region or city values of two tuples are similar, e.g., t_1 and t_2, then their corresponding addr or post values should also be similar.

Special Case: NEDs

Figure 1.1 shows that NEDs in Sect. 3.2 are special cases of CDs. When the similarity functions in a CD are defined on attributes in one table, it is exactly an NED. For instance, we can represent the example ned_1 in Sect. 3.2 by a CD,

$$cd_2 : \theta(name), \theta(address) \rightarrow \theta(street),$$

where

$$\theta(name) : [name \approx_{\leq 1} name],$$
$$\theta(address) : [address \approx_{\leq 5} address],$$
$$\theta(street) : [street \approx_{\leq 5} street].$$

It states that if the values of name and address are similar, then their corresponding street values should be similar as well. Therefore, CDs subsume the semantics of NEDs, or CDs generalize/extend NEDs, denoted by the arrow from NEDs to CDs in Fig. 1.1.

Discovery

Song et al. [152] introduce a pay-as-you-go approach for discovering comparable dependencies in a given dataspace. The algorithm is conducted in an incremental way with respect to new identified comparison functions, i.e., given a set of currently discovered dependencies and a newly identified comparison functions $\theta(A_i, A_j)$, it generates new dependencies w.r.t. $\theta(A_i, A_j)$.

For CDs discovery, both the error validation problem to determine whether $g_3 \leq e$, and the confidence validation problem to determine whether $conf \geq c$, are NP-complete [151], where g_3 measures the minimum number of tuples that have to be removed for the dependencies to hold, and $conf$ evaluates the maximum number of tuples such that the dependencies hold.

Application

Various applications of CDs are studied in [152]. The most important application is to improve the dataspace query efficiency. According to the comparison functions, the query evaluation searches not only the given attributes in a query tuple, e.g., region, but also their comparable attributes such as city. According to the comparable dependencies, if LHS attributes of the query tuple and a data tuple are found comparable, then the data tuple can be returned without evaluating the RHS attributes. It thus improves the query efficiency. In addition, CDs can be also used to improve data quality, such as detecting violation on heterogeneous data, and identifying duplicate tuples from various data sources.

3.6   Probabilistic Approximate Constraints (PACs)

Traditional FDs are rigid and limited to the subtle data quality problems caused by network data, such as existing problems changing with the network dynamics, new problems emerging over time and poor quality data. Therefore, there should be user-specified rule templates with open parameters for tolerance and likelihood. The new method can use statistical techniques to instantiate suitable parameter values from data and show how to apply them for monitoring data quality. Similar to PFDs/AFDs introducing probabilities and approximations to FDs in Sects. 2.4 and 2.5, Korn et al. [111] study probabilistic approximate constraints (PACs), bringing together distance metrics and probabilities. That is, PACs introduce tolerance and confidence factors into integrity constraints.

Definition

A *probabilistic approximate constraint* (PAC)

$$\text{PAC} : X_\Delta \to^\delta Y_\epsilon$$

specifics that, if two tuples have distances on attributes X as

$$|t_i[A_l] - t_j[A_l]| \leq \Delta_l, \forall A_l \in X,$$

then their probability of distances on attributes Y should be

$$\Pr(|t_i[B_l] - t_j[B_l]| \leq \epsilon_l) \geq \delta, \forall B_l \in Y,$$

where $|t_i[A_l] - t_j[A_l]|$ denotes the distance between tuples t_i and t_j on attribute A_l, $|t_i[B_l] - t_j[B_l]|$ is the distance on attribute B_l, Δ_l and ϵ_l are distance tolerances on attributes A_l and B_l, respectively, and δ is a confidence requirement.

Example

Consider again the example $r_{3.2}$ in Table 3.2.

The PAC with $\delta = 0.9$ below can tolerate that there are 10% of tuples not satisfying the distance constraints

$$\text{pac}_1 : \text{price}_{100} \to^{0.9} \text{tax}_{10}.$$

Table 3.2 does not satisfy this PAC. There are 11 tuple pairs that have price distances less than or equal to 100. Among them, 3 pairs of tuples have tax distances greater than 10. It follows $\Pr(|t_i[\text{tax}] - t_j[\text{tax}]| \leq 10) = 8/11 = 0.727 < \delta$.

Special Case: NEDs

PACs extend NEDs in Sect. 3.2 as presented in Fig. 1.1. When the probability threshold in a PAC is $\delta = 1$, it is exactly an NED. In other words, all NEDs can be represented as special PACs with $\delta = 1$. The example ned_1 in Sect. 3.2 can be expressed as a PAC as follows:

$$\text{pac}_2 : \text{name}_1 \text{address}_5 \to^1 \text{street}_5.$$

It states that when two tuples have similar names and addresses, and the streets should also be similar. And $\delta = 1$ means that we don't tolerate any tuples that do not satisfy the constraints. Consequently, PACs subsume the semantics of NEDs, or PACs generalize/extend NEDs, denoted by the arrow from NEDs to PACs in Fig. 1.1.

Discovery

The PAC Manager, namely PAC-Man in [111], provides a method to specify PACs. There are several parameters that need to be determined in a PAC, including Δ, ϵ for approximation with distances, and δ for the probability of satisfying the constraint. Given a set of rule templates provided by users and some training data, PAC-Man first instantiates the aforesaid parameters. Moreover, it keeps on monitoring the new data overtime and alarms when violations are detected.

Application

PAC-Man is integrated into a database for various applications [111]. It manages PACs for the entire aggregate network traffic database. The system can improve the data quality by picking thresholds suitably for the PACs, and monitoring new data in the database to check whether the data deviate from the PACs. It works as a browser or monitor that keeps on tracking data quality. When problems are detected, e.g., with missing values, instead of data cleaning, PAC-Man proposes to automatically rewrite users' queries over the remaining complete observations.

3.7   Fuzzy Functional Dependencies (FFDs)

Relational data models usually deal with well-defined data. However, in practical applications, data are usually partially known (incomplete) or imprecise. Through the proper interpretation of fuzzy membership function, fuzzy relational data model can be used to express the fuzziness of data values and the inaccuracy of their correlation, so as to meet the needs of the wider real world and provide closer human-computer interaction. Fuzzy relational data model has been proposed to represent ambiguities in data values as well as impreciseness in the association among them. Since there is no clear way of checking whether two imprecise values are equal, the traditional functional dependencies may not be directly applicable to the fuzzy relation schema. Therefore, functional dependencies have been extended [138] by replacing the equality comparison on domain values with "approximately equal", "more or less equal", etc.

Definition

For the domain of each attribute A_i, $dom(A_i)$, a fuzzy resemblance relation EQUAL $\mu_{EQ}(a, b), a, b \in dom(A_i)$, is defined to compare the elements of the domain, e.g., within the range of $[0, 1]$. It should be appropriately selected during database creation to capture the meaning of equality, or approximate equality, of domain values. For instance, the more the values a and b are "equal", the larger the $\mu_{EQ}(a, b)$ is (see example below). The fuzzy relation EQUAL is then extended over all attributes in R of tuples t_1 and t_2,

$$\mu_{EQ}(t_1, t_2) = \min\{\mu_{EQ}(t_1[A_1], t_2[A_1]), \mu_{EQ}(t_1[A_2], t_2[A_2]),$$
$$\ldots, \mu_{EQ}(t_1[A_n], t_2[A_n])\}.$$

A *fuzzy functional dependency* (FFD) has the form of

$$FFD : X \rightsquigarrow Y,$$

with $X, Y \subseteq R$. It holds in a fuzzy relation instance r on R, if for all tuples t_1 and t_2 of r, we have

$$\mu_{EQ}(t_1[X], t_2[X]) \leq \mu_{EQ}(t_1[Y], t_2[Y]).$$

Here, $\mu_{EQ}(t_1[X], t_2[X])$ is the fuzzy resemblance relation EQUAL of tuples t_1 and t_2 on attributes X. And $\leq$ means that the resemblance relation EQUAL of tuples t_1 and t_2 on attributes X is less than that of Y. It denotes that the values on attributes Y are more "equal" than those on attributes X.

Example

For the relation $r_{3.2}$ in Table 3.2, we consider an FFD

$$\text{ffd}_1 : \text{name, price} \rightsquigarrow \text{tax,}$$

where EQUAL is defined as follows

$$\mu_{EQ}(a, b) = \begin{cases} 0 & \text{if } a \neq b \\ 1 & \text{if } a = b \end{cases}, a, b \in dom(\text{name});$$

$$\mu_{EQ}(a, b) = 1/(1 + \beta|a - b|), \text{ where}$$

$$\beta = \begin{cases} 1 & \text{if } a, b \in dom(\text{price}) \\ 10 & \text{if } a, b \in dom(\text{tax}). \end{cases}$$

It states that for any two tuples having "equal" price should have "equal" tax.
   Consider two tuples t_1 and t_2 in Table 3.2. The EQUAL function is computed by

$$\mu_{EQ}(\text{NC, NC}) = 1, \quad \text{NC} \in dom(\text{name});$$

$$\mu_{EQ}(299, 300) = \frac{1}{1 + |299 - 300|} = \frac{1}{2}, \ 299, 300 \in dom(\text{price});$$

$$\mu_{EQ}(29, 20) = \frac{1}{1 + 10 \times |29 - 20|} = \frac{1}{91}, \ 29, 20 \in dom(\text{tax}).$$

It follows

$$\min(\mu_{EQ}(\text{NC, NC}), \mu_{EQ}(299, 300)) > \mu_{EQ}(29, 20).$$

That is, they conflict with the fuzzy functional dependencies, since

$$\min(\mu_{EQ}(t_1[\text{name}], t_2[\text{name}]), \mu_{EQ}(t_1[\text{price}], t_2[\text{price}]))$$
$$> \mu_{EQ}(t_1[\text{tax}], t_2[\text{tax}]).$$

Special Case: FDs

As shown in Fig. 1.1, FFDs extend FDs. For the example fd_1 in Sect. 1.1, by defining EQUAL as

$$\mu_{EQ}(a, b) = \begin{cases} 0 & \text{if } a \neq b \\ 1 & \text{if } a = b \end{cases}, a, b \in dom(\text{address}),$$

$$\mu_{EQ}(a, b) = \begin{cases} 0 & \text{if } a \neq b \\ 1 & \text{if } a = b \end{cases}, a, b \in dom(\text{region}),$$

we can obtain a corresponding FFD,

$$ffd_2 : \text{address} \rightsquigarrow \text{region}.$$

It states that if two tuples t_1 and t_2 have the same address, i.e.,

$$\mu_{EQ}(t_1[\text{address}], t_2[\text{address}]) = 1,$$

they must have the same region, i.e.,

$$\mu_{EQ}(t_1[\text{address}], t_2[\text{address}]) = 1 \leq \mu_{EQ}(t_1[\text{region}], t_2[\text{region}]).$$

Therefore, FFDs subsume the semantics of FDs, or FFDs generalize/extend FDs, denoted by the arrow from FDs to FFDs in Fig. 1.1.

Discovery

Wang and Chen [173] propose a mining algorithm for FFDs, which extends the TANE algorithm [95, 96] for FDs discovery as introduced in Sect. 2.1. Similar to the small-to-large search strategy in TANE (only for non-trivial FDs), it finds each non-trivial FFD which has a single attribute on its right-hand side. The algorithm considers every tuple pair to check whether it satisfies the EQUAL relation. Moreover, an incremental searching algorithm is further devised based on the pair-wise comparison [172]. When a new tuple is added, it avoids re-scanning the database.

Application

Analogous to FDs to identify the occurrence of redundancy in a database, FFDs are also used for redundancy elimination [24], with tolerance to slight difference in the data. Intan and Mukaidono [99] study query processing in a fuzzy relation in the presence of FFDs. Approximate join over multiple fuzzy relations is also introduced. Ma et al. [127] investigate the strategies and approaches for compressing fuzzy values by FFDs. The idea is to eliminate the unnecessary elements from a fuzzy value and thus compress its range.

3.8   Ontology Functional Dependencies (ONFDs)

Similar to MFDs, which extend traditional FDs with small changes in data format and interpretation, values with different syntax but the same semantics will be marked as data errors. Therefore, ontology functional dependencies (ONFDs) [9] are proposed to enhance dependencies-based data cleaning. Ontology FDs propose to represent the semantic attribute relationships, e.g., synonyms and is-a hierarchy defined by ontology.

Definition
Given a relation r and a set of attributes X, let x_i denote a set of tuples, which share the equal values in X. The index i is the smallest tuple index in x_i. $\prod_X$ is the set of equivalence classes. For instance, in Table 3.3, we have $\prod_{univ} = \{\{t_1, t_3\}, \{t_2, t_4\}\}$.

An ontology S represents how the categories form a domain, and the common categories include entities, concepts, relations and properties. Given a class E in S, synonyms(E) is the set of all synonyms for E. For example, synonyms$(E1) = \{\text{"BJ", "AAPL"}\}$, synonyms$(E2) = \{\text{"BJ", "Beijing"}\}$. name$(C)$ denotes the union of all classes. For instance, name$(\text{BJ}) = \{E1, E2\}$ as "BJ" can be a public company or a city. descendants(E) denotes a set of representations for E and its descendants.

ONFDs can be categorized into two sub-categories: synonym ONFDs and inheritance ONFDs.

A *synonym ontology functional dependency* has the form as follows:

$$\text{synonyms ONFD} : X \rightarrow_{syn} A,$$

where syn means that for each equivalent class $x \in \prod_X$, there exists at least one interpretation under which all the A-values of x are synonyms, i.e.,

$$\bigcap_{names(a),a\in\{t[A]|t\in x\}} \neq \emptyset.$$

An *inheritance ontology functional dependency* has the form as follows,

$$\text{inheritance ONFD} : X \rightarrow_{inh} A,$$

Table 3.3 An example relation instance $r_{3.3}$ of faculties

	Name	Univ	City	Dept	Email
t_1	Dong Li	HKUST	HK	CS	dl@cse.ust.hk
t_2	Richard Wang	THU	Beijing	Physics	rw@tsinghua.edu.cn
t_3	D. Li	HKUST	Hong Kong	Information	dl@cse.ust.hk
t_4	Richard Wang	THU	BJ	Science	rw@tsinghua.edu.cn

where inh means that for each equivalent class $x \in \prod_X$, the A-values of all tuples $t_i \in x$ are descendants of a Least Common Ancestor (LCA) which is within a distance of θ in S, i.e.,

$$\bigcap_{\text{descendant}(a), a \in \{t[A] | t \in x\}} \neq \emptyset.$$

Example

For the relation $r_{3.3}$ in Table 3.3, there is a synonyms ONFD as follows,

$$\text{onfd}_1 : [\text{univ}] \to_{\text{syn}} [\text{city}].$$

In this example, we have $\prod_{\text{univ}} = \{\{t_1, t_3\}, \{t_2, t_4\}\}$. The first equivalent class $\{t_1, t_3\}$ represents the university "HKUST". According to the city ontology, names("BJ") $\cap$ names("Beijing") = "Beijing". Consider an inheritance ONFD from Table 3.3 and Fig. 3.1,

$$\text{onfd}_2 : [\text{email}] \to_{\text{inh}} [\text{dept}].$$

The first equivalent class $\{t_1, t_3\}$ represents the email address "dl@cse.ust.hk". The LCA is "Information" which is within a distance of one ($\theta = 1$) to each dept value in this class.

Special Case: FDs

As shown in Fig. 1.1, ONFDs subsume FDs in Sect. 1.1. Continued with example fd$_1$ in Sect. 1.1, it can be rewritten as an ONFD as

$$\text{onfd}_3 : [\text{address}] \to_{\text{syn}} [\text{region}],$$

where all values are assumed to have a single literal interpretation, i.e., for all classes E, $|synonyms(E)| = 1|$. To be specific, for the relation instance $r_{1.1}$ in Table 1.1, we have $\prod_{\text{address}} = \{\{t_1, t_2\}\{t_3, t_4\}\{t_5, t_6\}\{t_7\}\{t_8\}\}$. If for all classes E, $|synonyms(E)| = 1|$, there is only "New York" for $\{t_1, t_2\}$, and similarly there is only "Boston, MA" for $\{t_7\}$, so on and so forth. However, there are "Boston" and "Chicago, MA" for $\{t_3, t_4\}$, i.e., violating onfd$_3$. Nevertheless, ONFDs subsume FDs, or ONFDs generalize/extend FDs, denoted by the arrow from FDs to ONFDs in Fig. 1.1.

Fig. 3.1 An example department ontology

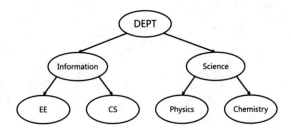

Discovery

Baskaran et al. [9] propose an algorithm for discovering complete and minimal set of synonym and inheritance ONFDs. The algorithm prunes the prior search space using exact axioms and axioms with some exceptions. The set of possible antecedent and consequent values can be modeled as a set containment lattice. The algorithm firstly searches singleton sets of attributes, then it searches for larger attribute sets level by level through the lattice. ONFDs can be found by traversing a set inclusion lattice with exponential worst-case complexity.

Application

As mentioned in [9], ONFDs can be used for data cleaning. They provide rich context information to distinguish synonymous values from inherited values, which are not syntactically equivalent but refer to the same entity. ONFDs can significantly reduce false positive errors in data cleaning techniques. In addition, ONFDs can identify potential dirty data values and provide suggestions for possible correct values. In this sense, as a new data quality rule to capture domain relationships, ONFDs with higher support contain more reliable values to repair consistent and potentially dirty values.

3.9 Matching Dependencies (MDs)

In record matching [54], the matching rules have to adapt to errors and different representations in different data sources. In this case, the matching rules could also be interpreted as data dependencies: if some attributes match then identify other attributes.

Matching dependencies (MDs) [62, 70] specify matching rules for object identification. Bertossi et al. [20] study the semantics of MDs.

Unlike traditional dependencies, MDs have dynamic semantics to accommodate errors in unreliable data sources. They are defined by similarity operators, spanning potentially different relationships. In other words, MDs consider similarity metrics on determinant attributes X to determine the identification of dependent attributes Y.

Definition

A *matching dependency* (MD) on a relation R has a form[1]

$$\text{MD} : X \approx \rightarrow Y \rightleftharpoons,$$

where (1) $X \subseteq R$, $Y \subseteq R$; (2) $\approx$ denotes the corresponding *similarity operator* on attributes of X, which indicates that two values are similar; (3) $\rightleftharpoons$ denotes the *matching operator* on

---

[1] Similar to CDs in Sect. 3.5, MDs can also be defined on attributes of two relations from heterogeneous sources [62, 70].

attributes of Y, which shows that two values are identified. It states that for any two tuples from an instance of relation R, if they are similar on attributes X, then their Y values should be identified.

Example

Consider the relation $r_{3.2}$ in Table 3.2, An MD can be

$$md_1 : \text{street} \approx, \text{region} \approx\rightarrow \text{zip} \rightleftharpoons .$$

It states that if any two tuples, e.g., t_5 and t_6 from $r_{3.2}$ in Table 3.2, have similar streets (with edit distance ≤ 5 denoted by $\approx$) and similar regions (with distance ≤ 2 denoted by $\approx$), then they can be identified (denoted by $\rightleftharpoons$) on zip.

Special Case: FDs

Figure 1.1 states that MDs subsume FDs in Sect. 1.1. When the values have matching similarity equal to 1.0, it is exactly an FD. The example fd_1 in Sect. 1.1 can thus be represented by a special MD

$$md_2 : \text{address} =\rightarrow \text{region} =,$$

when address and region have identical values, respectively. In this sense, MDs subsume the semantics of FDs, or MDs generalize/extend FDs, denoted by the arrow from FDs to MDs in Fig. 1.1.

Discovery

Song and Chen [145, 147] propose both exact and approximation algorithms for discovering MDs. The exact algorithm traverses all the data to find MDs that satisfy the required confidence and support, i.e., evaluations of MDs.

The approximation algorithm only traverses the first k tuples in statistical distribution, with bounded relative errors on support and confidence of returned MDs. Moreover, similar to minimal/candidate keys about FDs, relative candidate keys (RCKs) with minimal compared attributes can remove redundant semantics. Song et al. [150] find a concise set of matching keys, which can reduce the redundancy while satisfying the coverage and validity.

The problem of deciding whether there exists a set of matching keys such that $supp \geq s$, $conf \geq c$, and the size of the set is no greater than k is NP-complete [150], where $supp$ measures the proportion of distinct tuple pairs that agree on at least one of the matching keys in the set, and $conf$ is the minimum ratio of tuple pairs that satisfies a matching key in the set.

Application

Record linkage, also known as entity identification or reference reconciliation, aims to identify the duplicate records in a database (see [54] for a survey). MDs have been successfully utilized as matching rules in detecting duplicates. Similarity metrics and dynamic semantics help largely in addressing the unreliability of data.

Reasoning mechanism for deducing MDs from a set of given MDs is studied in [70]. MDs and their reasoning techniques can improve both the quality and efficiency of various record matching methods. With a comprehensive inference mechanism for MDs, some duplicates, which cannot be identified by a given set of MDs, could be detected by the inferred dependencies.

Remarkably, record matching with MDs and data repairing with CFDs can interactively perform together [72, 76]. While matching aims to identify tuples that refer to the same real-world object, repairing is to make a database consistent by fixing data errors under integrity constraints. The interaction between record matching and data repairing can effectively help each other.

3.10 Conditional Matching Dependencies (CMDs)

In practice, MDs may be too strict for all the tuples in a relation. Therefore, some data dependencies that only refer to a part of a table can be applied conditionally to a subset of tuples. Again, analogous to CFDs extending FDs in Sect. 2.7, Wang et al. [175] study conditioning MDs in a subset of tuples, namely conditional matching dependencies (CMDs). The idea is to bind matching dependencies to only a portion of the table. Compared with MDs, the expression ability of CMDs is stronger, which can meet the needs of a wider range of applications.

Definition

A *conditional matching dependency* (CMD) has a form

$$CMD : X \approx x \rightarrow Y \rightleftharpoons y,$$

where x and y are domain values of attributes $X \subseteq R$ and $Y \subseteq R$, i.e., $x \in dom(X)$ and $y \in dom(Y)$, respectively, or a virtual value $*$ that is similar/identified with any value. It states that for any two tuples from an instance of relation R, if they are similar on attributes X and similar to x as well, then their Y values should be identified with each other and also identified with y.

Example

A CMD example on the relation $r_{3.2}$ in Table 3.2 can be

$$cmd_1 : street \approx Central\ Park, region \approx * \rightarrow zip \rightleftharpoons *.$$

It states that only for the tuples whose street values are similar with each other and also similar to "Central Park", e.g., the tuples t_1 and t_3 in Table 3.2, if their region values are also similar, they must have identified zip. Since the virtual value $*$ is similar to any value, region $\approx *$ only requires two tuples to have similar region values, i.e., no condition is specified on region. This CMD does not take effect over other subsets of tuples whose street is not similar to the condition of "Central Park".

Special Case: MDs

As shown in Fig. 1.1, MDs in Sect. 3.9 are special cases of CMDs. When the values of X match for the identifications of Y without conditions, it is exactly an MD. For example, md_1 in Sect. 3.9 can be represented as a CMD,

$$\mathsf{cmd}_2 : \mathsf{street} \approx *, \mathsf{region} \approx * \rightarrow \mathsf{zip} \rightleftharpoons *,$$

where only $*$ appears without a constant. Hence, CMDs subsume the semantics of MDs, or CMDs generalize/extend MDs, denoted by the arrow from MDs to CMDs in Fig. 1.1.

Discovery

Similar to the discovery of MDs in Sect. 3.9, redundant semantics exist among CMDs and should be removed in discovery. Wang et al. [175] define irreducible CMDs, which determine the target $Y \rightleftharpoons y$ and are redundancy free. Analogous to CFD discovery in Sect. 2.7, several strategies are studied for CMD discovery, e.g., a domain-oriented algorithm by enumerating attributes and values or a tuple-oriented algorithm by traversing and pruning tuple pairs. The complexity of the discovery is $O(n^2|dom(A)|^m)$, where n is the number of tuples in a relation, m is the number of attributes and A is an attribute.

The problem of deciding whether a CMD has $g_3 \leq e$ in a relation is NP-complete [175], where g_3 is the error rate of the CMD, the minimum number of tuples that need to be removed from the relation in order to make the CMD hold, and e is the maximum bound of error rate.

Straight-forward Algorithm is to evaluate all the possible candidates of the search space. It adds all the CMDs that hold the instances and finally removes reducible ones. *Domain-oriented Algorithm* utilizes the left-dominating inference rule for pruning CMDs. *Tuple-oriented Algorithm*, which avoids traversing all the tuples in each evaluation, incrementally excludes tuple pairs with respect to the given instance and introduces dynamic domain according to the currently remaining tuple pairs.

Application

Same as MDs, CMDs can be used as matching rules in a rule-based method of record linkage [175]. It tells how the attributes in a relation should be compared in order to identify duplicate records. Moreover, the conditions in CMDs further recognize the subset of tuples where the matching rules are applicable.

3.11   Summary and Discussion

To capture the relationships among heterogeneous data, instead of the equality operator in FDs over categorical data, distance and similarity of values are considered. The distance/similarity metrics could be introduced in the left-hand side, right-hand side or both sides of attributes in a data dependency. They can also be extended to two relations from heterogeneous sources [151]. Such extensions on distance and similarity are useful not only for violation detection but also object identification [70]. Similarly, the distance metrics could also be considered over numerical data, e.g., in sequential dependencies (SDs) [87] in Sect. 4.5. Referring to Table 3.1 and Fig. 3.2, we summarize the relationships among the extensions over heterogeneous data in different (sub)categories as follows. Table 3.4 compares their major features in extension.

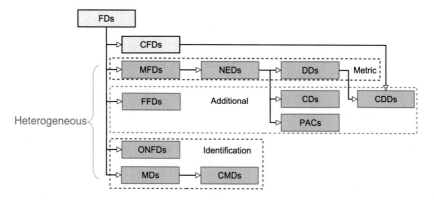

Fig. 3.2 A family tree of extensions between heterogeneous data dependencies

Table 3.4 Features of extensions over heterogeneous data

Subcategory	Sections	Dependency	Determinant X	Dependent Y
Metric	3.1	MFDs	Equality	Distance
Metric	3.2	NEDs	Distance	Distance
Metric	3.3	DDs	Distance	Distance
Additional-condition	3.4	CDDs	Distance	Distance
Additional-attribute	3.5	CDs	Distance	Distance
Additional-probability	3.6	PACs	Distance	Distance
Additional-fuzzy	3.7	FFDs	Distance	Distance
Identification	3.8	ONFDs	Equality	Synonym
Identification	3.9	MDs	Similarity	Identification
Identification	3.10	CMDs	Similarity	Identification

Metric

While MFDs [112] introduce a distance metric on the dependent attribute Y, NEDs [10] employ the distance metric in both the determinant X and dependent Y. In addition to the similarity relationship, DDs [146] further study the dissimilarity using the distance metric. While the extensions with metrics enable data dependencies with more expressive power and tolerance to variations of heterogeneous data, the data dependencies become more complicated. In particular, the thresholds of metric distances/similarities are often non-trivial to specify in practice.

Additional

To address more complicated scenarios, additional extensions are introduced to the distance metric. Analogous to CFDs extending FDs in Sect. 2.7, CDDs [116] also extend DDs with conditions, i.e., DDs hold only in a subset of tuples specified by the conditions. In addition to the similarity on heterogeneous values, CDs [151, 152] further consider the similarity on attribute names of relations from heterogeneous sources. Finally, similar to PFDs introducing probability to FDs in Sect. 2.4, PACs [111] bring together distance metrics and probability, i.e., introducing tolerance and confidence factors into integrity constraints. Ambiguities in heterogeneous data values make it difficult to check whether two values are equal. FFDs [138] thereby extend FDs by replacing the equality comparison on domain values with "approximately equal", "more or less equal", etc.

Identification

To capture the identification relationships, ONFDs [9] employ synonyms and is-a hierarchy defined by ontology. While ONFDs consider equal X values to imply the identification of Y, MDs [54] further consider similarity on X values similar to DDs or CDs. Again, analogous to CFDs extending FDs in Sect. 2.7, CMDs [175] study conditioning MDs in a subset of tuples.

Ordered Data

4

Data sets with ordered values are prevalent, e.g., timestamps, sequence numbers, sales, temperature, stock prices and so on. It is also promising to address data dependencies on such ordered data. In the following, we introduce several typical studies on such ordered data, where each attribute A has a partial ordering $\leq_A$ on $dom(A)$. Data dependencies are then declared concerning the orderings on the attribute domains [50, 83–85].

The Ordered part in Fig. 1.1 presents the major data dependencies extended for numeral data. The corresponding special cases, where the data dependencies are extended from, are presented as well. Each pair of a data dependency and its special case, e.g., SDs extend ODs, corresponds to an arrow from ODs to SDs in the family tree in Fig. 1.1.

Table 4.1 further categorizes the extensions in the following aspects. (1) Order: To express the constraints on orders, operators such as $<, \leq, >, \geq$ are employed. In particular, with further extensions on $=$ and $\neq$ operators, the data dependencies with orders are sufficient to capture the traditional semantics such as functional dependencies or conditional functional dependencies. (2) Distance: While the order-based extensions express the relationships on increasing or decreasing, they do not constrain the amount of value increase or decrease. To this end, analogous to differential dependencies, the distances of value changes with respect to ordering are specified.

4.1 Ordered Functional Dependencies (OFDs)

There is strong evidence that in many database applications, sorting is inherent in the underlying structure of data, and linear sorting is particularly important for advanced applications involving temporary oral or scientific information. Therefore, an ordered relational model can be defined. Ng [134, 135] considers data dependencies with the semantics of domain orderings. Intuitively, it specifies that values in attributes should be ordered similarly, for

© The Author(s), under exclusive license to Springer Nature Switzerland AG 2023   71
S. Song and L. Chen, *Integrity Constraints on Rich Data Types*,
Synthesis Lectures on Data Management,
https://doi.org/10.1007/978-3-031-27177-9_4

Table 4.1 Category and special cases of data dependencies over ordered data, where each pair corresponds to an arrow in Fig. 1.1 such as DCs◁–ODs

Subcategory	Sections	Data dependencies	Special cases
Order	4.1	OFDs	
Order	4.2	ODs	OFDs, FDs
Order	4.3	BODs	ODs
Order	4.3	DCs	ODs, eCFDs
Distance	4.5	SDs	ODs
Distance	4.6	CSDs	SDs

instance, the mileage increases with time. Two advanced domain orderings, i.e., pointwise-orderings and lexicographical orderings, are introduced. Intuitively, pointwise-orderings require each component of a data value to be greater than its predecessors and lexicographical orderings resemble the way in which words are arranged in a dictionary.

Definition

An *ordered functional dependency* (OFD) is classified into POFDs with pointwise-orderings and LOFDs with lexicographical orderings.

An *ordered functional dependency* arising from *pointwise-orderings* (POFD) over a relation R has the form

$$\text{POFD} : X \rightarrow^P Y.$$

Pointwise ordering on X, denoted by $t_1[X] \leq_X^P t_2[X]$, means that, for n attributes A_i in X, $1 \leq i \leq n, t_1[A_i] \leq t_2[A_i]$. The POFD states that for all tuples t_1, t_2 of relation $R, t_1[X] \leq_X^P t_2[X]$ implies that $t_1[Y] \leq_Y^P t_2[Y]$.

Similarly, there is also an *ordered functional dependency* arising from *lexicographical orderings* (LOFDs), which has the form of

$$\text{LOFD} : X \rightarrow^L Y.$$

Lexicographical ordering $t_1[X] \leq_X^L t_2[X]$ on X means that, for n attributes A_i in X, either (1) there exists k with $1 \leq k \leq n$ such that $t_1[A_k] < t_2[A_k]$, and for all $1 \leq i < k, t_1[A_i] = t_2[A_i]$; or (2) for all $1 \leq i \leq n, t_1[A_i] = t_2[A_i]$. Compared to POFDs, the number k enables LOFDs to express the constraint that some attributes in two tuples are equal.

Example

A POFD can be declared over $r_{4.2}$ in Table 4.2 as follows,

$$\text{pofd}_1 : \text{subtotal} \rightarrow^P \text{taxes}.$$

Table 4.2 Relation instance $r_{4.2}$ with multiple ordered attributes on hotel rates

	Nights	Unit-price	Subtotal	Taxes	Star
t_1	1	190	190	38	3
t_2	2	185	370	74	3
t_3	3	180	540	108	4
t_4	4	175	700	140	4

It states that a higher subtotal leads to higher taxes. For example, the subtotal of t_2 is "370", higher than that of t_1 "190". Thereby, the taxes of t_2 should be higher as well.

We can also set an LOFD as follows,

$$\text{lofd}_1 : \text{star, subtotal} \to^{\mathsf{L}} \text{tax},$$

which states that, the tax of a hotel is greater than that of any other hotel with the same star but lower subtotal or with the same star and the same subtotal. For example, with $k = 2$, t_1 and t_2 satisfy this constraint, because with the same star, $t_1[\text{subtotal}] = 190 < t_2[\text{subtotal}] = 370$ and $t_1[\text{taxes}] = 38 < t_2[\text{taxes}] = 74$. With $k = 1$, t_2 and t_3 also satisfy the constraint since $t_2[\text{star}] = 3 < t_3[\text{star}] = 4$, $t_2[\text{subtotal}] = 370 < t_3[\text{subtotal}] = 540$ and $t_2[\text{taxes}] = 74 < t_3[\text{taxes}] = 108$.

Axiomatization

For the class of ordered data, following Armstrong's axiom system, Ng [134] gives a set of inference rules for POFDs. The axiom system with four rules is complete and sound. Let $X, Y, Z, W \subseteq R$ and Σ be a set of POFDs over R. $X \sim Y$ means that two sequences $X = \langle A_1, ..., A_m \rangle$ and $Y = \langle B_1, ..., B_n \rangle$ have the same elements, i.e., $\{A_1, ..., A_m\} = \{B_1, ..., B_n\}$.

POFD1 (Reflexivity): If $Y \subseteq X$, then $\Sigma \vdash X \to^P Y$.
POFD2 (Augmentation): If $\Sigma \vdash X \to^P Y$, then $\Sigma \vdash XZ \to^P YZ$.
POFD3 (Transitivity): If $\Sigma \vdash X \to^P Y$ and $\Sigma \vdash Y \to^P Z$, then $\Sigma \vdash X \to^P Z$.
POFD4 (Permutation): If $\Sigma \vdash X \to^P Y$, $W \sim X$ and $Z \sim Y$, then $\Sigma \vdash W \to^P Z$.

POFD1, POFD2 and POFD3 are modified according to Armstrong's axioms and POFD4, dealing with sequence of attributes, is specific to POFDs.

Application

Ordering is inherent to the underlying structure of data in many database applications, such as textual and software information. It is found that monotonicity properties arise naturally in many applications, and OFDs can capture a monotonicity property between two sets of values projected onto some attributes in a relation. For example, OFDs can capture the constraint between the salary and experience or positions for an employee. In addition, OFDs can also be used in business-oriented applications such as accounting and payroll processing.

The appropriateness of the choice of the POFD or the LOFD in this case depends entirely on the semantics of the promotion policy adopted by the users.

OFDs are employed as useful semantic constraints over temporal relations [133]. Temporal relations can be interpreted as special cases of linearly ordered relations over time schemas. OFDs can hold the consistency of time data with various time measurement systems. For example, the experience of an employee should be increased with the passage of time.

4.2   Order Dependencies (ODs)

Rather than considering the attributes in the same increasing order as OFDs in Sect. 4.1, order dependencies (ODs) [50, 83–85] are introduced to express the constraints with different orders. For instance, the price of a production drops with the increase of time. ODs express semantic information concerning the orderings on the attribute domains of a relation, such as between timestamps and numbers, which are common in business data. Therefore, ODs improve the expression power of constraints and have been applied to enhance database implementation.

Definition

For each attribute A, the marked attributes of A are used to denote various orderings, such as $A^<$, $A^\leq$, $A^=$, $A^>$, $A^\geq$ and so on. For any two tuples t_1, t_2, $t_1[A^\leq]t_2$ means $t_1[A] \leq t_2[A]$. An *order dependency* (OD) over R is in the form of

$$\text{OD} : X \rightarrow Y,$$

where X and Y are marked attributes. A relation instance r over schema R satisfies the OD, if for any two tuples $t_1, t_2 \in r$, $t_1[X]t_2$ implies $t_1[Y]t_2$.

Example

For the relation instance $r_{4.2}$ in Table 4.2, an OD can be

$$\text{od}_1 : \text{nights}^\geq \rightarrow \text{unit-price}^\leq.$$

That is, the more nights a guest stays, the lower the average price per night (unit-price) will be. For instance, for the tuples t_1 and t_2, there is $t_1[\text{nights}] = 1 \leq 2 = t_2[\text{nights}]$. It leads to $t_1[\text{unit-price}] = 190 \geq 185 = t_2[\text{unit-price}]$.

Special Case: OFDs

Figure 1.1 indicates that ODs extend OFDs in Sect. 4.1. When all the marked attributes are in the form of $A^\leq$ in an OD, it is exactly an OFD. For example, pofd_1 in Sect. 4.1 can be represented as an OD,

$$\text{od}_2 : \text{subtotal}^\leq \rightarrow \text{taxes}^\leq.$$

It also means that the taxes of a tuple should be higher than that of any other tuples with less subtotal.

In this sense, ODs subsume the semantics of OFDs, or ODs generalize/extend OFDs, denoted by the arrow from OFDs to ODs in Fig. 1.1.

Special Case: FDs

Table 4.1 indicates that ODs extend FDs. When all the marked attributes are in the form of $A^=$ in an OD, it is exactly a FD. For example, fd : subtotal $\rightarrow$ taxes can be represented as an OD,

$$od_3 : \text{subtotal}^= \rightarrow \text{taxes}^=.$$

In this sense, ODs subsume FDs, denoted by the arrow from FDs to ODs in Fig. 1.1.

Axiomatization

For ODs, the axiom system is also complete and sound [83], consisting of seven inference rules. Let M, N, P, Q be sets of linear marked attributes, A be an individual attribute, $\hat{A}^R$ be the reversal of a marked attribute $\hat{A}$, M^R be the set $\{\hat{A}^R \mid \hat{A} \text{ in } M\}$, supp$(M)$ be the set $\{A \mid \text{some marked attribute of } A \text{ is in } M\}$.

OD1 (Reflexivity): If $N \subseteq M$, then $M \rightarrow N$.

OD2 (Augmentation): If $M \rightarrow N$ and $Q \subseteq P$, then $MP \rightarrow NQ$.

OD3 (Transitivity): If $M \rightarrow N$, $N \rightarrow P$, then $M \rightarrow P$.

OD4 (Reversal): If $M \rightarrow N$, then $M^R \rightarrow N^R$.

OD5 (Disjunction): If $MA^< \rightarrow N$ and $MA^= \rightarrow N$, then $MA^{<=} \rightarrow N$.

OD6 (Total order): If A has total order, $MA^< \rightarrow N$, $MA^= \rightarrow N$ and $MA^> \rightarrow N$, then $M \rightarrow N$.

OD7 (Impropriety): For each linear marked attribute $\hat{B}$, if $M = \text{supp}(M)$ and $M \rightarrow A^<$, then $\emptyset \rightarrow \hat{B}$.

OD1, OD2 and OD3 are analogous to Armstrong's axioms. OD$_4$ says that the reversal of an OD still holds. OD$_5$ can help us infer $A^\le$ from $A^<$ and $A^=$. OD$_6$ studies the feature of attributes with total order. OD$_7$ handles the cases where the original OD schema is improper.

Discovery

Langer and Naumann [117] propose an OD discovery algorithm, which traverses the lattice of permutations of attributes in a level-wise bottom-up manner. Instead of expressing ODs in a list notation, Szlichta et al. [160] express ODs with sets of attributes via a polynomial mapping into a set-based canonical form. Similar to the idea of FastFD [180], an algorithm named FASTOD is presented to discover a complete, minimal set of set-based ODs. It traverses a lattice of all possible sets of attributes in a level-wise manner, reducing the complexity of using list-based axiomatization. The implication problem for ODs, i.e., implying an OD from a set of ODs, is co-NP-complete [162].

Application

ODs are used for database implementation to improve efficiency [50]. Firstly, ODs can reduce indexing space without much access time increase. For instance, suppose that an employee database is sorted by rank. One can access the employee instance using the order of rank. In addition, if the database also satisfies the OD rank → salary, then the data is also ordered by salary. That is, we can get salary by rank rapidly.

To give another example in [50], if the file CHKS contains the contents of the instance CHECKS in order of CHECK#, a sparse index can be used to gain rapid access into CHKS via CHECK# value. Also, because CHECKS satisfies the order dependencies CHECK# → DATE, the file CHKS is ordered by DATE. Thus a sparse index can be used on DATE as well. This will yield a considerable space saving over a dense index, without increasing access times.

Moreover, ODs can also be used in query optimization. Szlichta et al. [161] present optimization techniques using ODs for various SQL functions and algebraic expressions. Finally, following the same line of other data dependencies, ODs are naturally applicable as integrity constraints for error detection and data repairing.

4.3   Band Order Dependencies (BODs)

Compared to ODs, band order dependencies, a.k.a. Band ODs, allow order relationships between attributes to contain small variations but still preserve semantics [119]. In many real-world scenarios, this makes sense. For instance, the release dates of music records almost increase as the catalog numbers increase. However, in the music industry, catalog numbers are often assigned to a record before its actual release date. It means that tuples with delayed release dates may slightly violate the OD between "catalog" and "release date". To capture the semantics, Li et al. [119] propose to further consider a permissible range (band) for order relationships.

Definition

Given a band width Δ, a list of attributes X over a certain relation R, for any two tuples $t, s \in R$, $t \preceq_{\Delta, X} s$ means that $d(s.Y, t.Y) \leq \Delta$, where $d(x_1, x_2) = ||x_2|| - ||x_1||$ and $||x||$ denotes the norm of the value list x. Let $t \preceq_X s$ be the operator $t \preceq_{\Delta, X} s$, where $\Delta = 0$.

A *band order dependency* (BOD) over R is in the form of

$$X \mapsto_\Delta Y,$$

where X and Y are marked attributes. A relation instance r over schema R satisfies the BOD, if for any two tuples $t, s \in r$, $t \preceq_X s$ implies $t \preceq_{\Delta, Y} s$.

Example

Consider the relation instance $r_{4.3}$ in Table 4.3. We have

$$BOD_1 : cat\# \mapsto_{\Delta=1} year.$$

That is, for two records t_i and t_j, t_i can be published at most one year earlier than t_j, if t_i has a catalog number greater than t_j. For instance, for the tuples t_3 and t_2, we have $t_3[cat\#] > t_2[cat\#]$. It leads to $t_2[year] - t_3[year] \leq \Delta = 1$.

Special Case: ODs

Table 4.1 indicates that band ODs extend ODs. When $\Delta = 0$ in a BOD, it is exactly an OD. For example, if we set $\Delta = 0$ in BOD_1, it will obtain BOD_2,

$$BOD_2 : cat\# \mapsto_{\Delta=0} year.$$

It means t_i cannot be published earlier than t_j, if t_i has a catalog number greater than t_j. It can also be represented as an OD, i.e., od : $cat\#^{\geq} \rightarrow year^{\geq}$.

Discovery

Li et al. [119] propose a band OD discovery algorithm. Given the band width Δ and attributes X and Y, the algorithm calculates the longest monotonic band (LMB) to verify whether a band OD holds. To match real scenarios, Li et al. [119] allow band ODs to hold approximately with some exceptions by considering an approximation ratio. The ratio indicates the minimum proportion of tuples that have to be removed to ensure the band OD holds.

Application

BODs are used for identifying data quality errors [119]. Given a BOD, the clean tuples are labeled by the longest monotonic band (LMB), while other tuples are considered to contain incorrect values. For instance, Li et al. [119] utilize BODs to detect outliers on two real-world datasets. In music industry, the discovered BODs find that the catalog numbers of music records may be assigned before the production year. This is because the production time of a record can range from a short period of time to a few years. BODs can capture this semantic and still identify outliers with LMB.

Table 4.3 An example relation instance $r_{4.3}$ of music records

	Year	Cat#
t_1	1999	47282
t_2	2000	47383
t_3	1999	47388

4.4   Denial Constraints (DCs)

Rather OFDs and ODs which only use $<, \leq, >, \geq$ to express orders without specifying concrete ranges, denial constraints (DCs) [17, 18] study a more general form of integrity constraints in addition to the dependencies between determinant and dependent attributes. It expresses the restrictions that prevent some attributes from taking certain values, by built-in atoms with $\{=, \neq, <, >, \leq, \geq\}$. The general form studied in [38] can further capture the semantics of other data dependencies such as functional dependencies (FDs).

Definition

A *denial constraint* (DC) has a form

$$DC : \forall t_\alpha, t_\beta, \cdots \in R, \neg(P_1 \wedge \cdots \wedge P_m),$$

where P_i is of the form $v_1 \phi v_2$ or $v_1 \phi c$, ϕ is an element of a negation closed finite operator set $\{=, \neq, <, >, \leq, \geq\}$, and $v_1, v_2 \in t_\alpha.A, t_\beta.A, \ldots, A \in R$, and c is a constant. It states that all the predicates cannot be true at the same time.

Example

Consider the relation instance $r_{4.2}$ in Table 4.2. The order relationships on numeric values such as subtotal and taxes can be captured by a DC with operators $<, >$ as follows,

$$dc_1 : \forall t_\alpha, t_\beta \in R, \neg(t_\alpha.\mathsf{subtotal} < t_\beta.\mathsf{subtotal}$$

$$\wedge t_\alpha.\mathsf{taxes} > t_\beta.\mathsf{taxes}).$$

It declares that a lower subtotal should not pay more taxes. For example, tuples t_1 and t_2 satisfy dc_1, having $t_1[\mathsf{subtotal}] < t_2[\mathsf{subtotal}]$ and $t_1[\mathsf{taxes}] < t_2[\mathsf{taxes}]$.

Special Case: ODs

Figure 1.1 states that DCs subsume ODs in Sect. 4.2. We can represent od_1 in Sect. 4.2 by a DC as follows:

$$dc_2 : \forall t_\alpha, t_\beta \in R,$$

$$\neg(t_\alpha.\mathsf{nights} \geq t_\beta.\mathsf{nights} \wedge t_\alpha.\mathsf{unit\text{-}price} > t_\beta.\mathsf{unit\text{-}price}).$$

Both constraints state that any $t_\alpha, t_\beta \in R$ should *not* have $t_\alpha[\mathsf{nights}] \geq t_\beta[\mathsf{nights}]$ but $t_\alpha[\mathsf{unit\text{-}price}] > t_\beta[\mathsf{unit\text{-}price}]$, i.e., stays more nights but with a higher unit-price. In this sense, DCs subsume the semantics of ODs, or DCs generalize/extend ODs, denoted by the arrow from FDs to DCs in Fig. 1.1.

Special Case: eCFDs

Again, Fig. 1.1 shows that DCs extend eCFDs. We can represent the example ecfd_1 in Sect. 2.8 by a DC,

$$\mathsf{dc}_3 : \forall t_\alpha, t_\beta \in R,$$
$$\neg(t_\alpha.\mathsf{rate} = t_\beta.\mathsf{rate} \wedge t_\alpha.\mathsf{rate} \leq 200 \wedge t_\alpha.\mathsf{name} = t_\beta.\mathsf{name}$$
$$\wedge\, t_\alpha.\mathsf{address} \neq t_\beta.\mathsf{address}).$$

It states that for any $t_\alpha, t_\beta \in R$, they should *not* have the same rate (≤ 200) and the same name but different addresses. Consequently, DCs subsume the semantics of eCFDs, or DCs generalize/extend eCFDs, denoted by the arrow from eCFDs to DCs in Fig. 1.1.

Axiomatization

Chu et al. [38] present three inference rules for DCs, including Triviality, Augmentation and Transitivity, and prove that the inference system is sound and also complete for a form of DCs. Let P and Q be predicates, $\overline{P}$ be the inverse of P, $\mathsf{Imp}(P)$ be the set of implied predicates of P.

DC1 (Triviality): $\forall P_i, P_j$, if $\overline{P_i} \in \mathsf{Imp}\left(P_j\right)$, then $\neg\left(P_i \wedge P_j\right)$ is a trivial DC.

DC2 (Augmentation): If $\neg(P_1 \wedge \ldots \wedge P_n)$ is a valid DC, then $\neg(P_1 \wedge \ldots \wedge P_n \wedge Q)$ is also a valid DC.

DC3 (Transitivity): If $\neg(P_1 \wedge \ldots \wedge P_n \wedge Q_1)$ and $\neg(R_1 \wedge \ldots \wedge R_m \wedge Q_2)$ are valid DCs, and $Q_2 \in \mathsf{Imp}\left(\overline{Q_1}\right)$, then $\neg(P_1 \wedge \ldots \wedge P_n \wedge R_1 \wedge \ldots \wedge R_m)$ is also a valid DC.

All three inference rules can find a counterpart in Armstrong's axioms. If DCs contain two predicates that cannot be true at the same time, DC1 would identify these DCs as trivial DCs. DC2 means that adding more predicates to DCs will result in valid DCs. DC3 states that if two valid DCs contain two predicates that cannot be false at the same time, then we can generate a new DC by removing the two predicates.

Discovery

Chu et al. [38] present a DC discovery algorithm FASTDC as an extension of FastFD [180]. It starts by building a predicate space and calculates evidence sets. For instance, for the DCs involving at most two tuples without constants, the structure of a predicate consists of two different attributes and one operator. The algorithm establishes the connection between discovering minimal DCs and finding minimal set covers for evidence sets. Depth-first search strategy is deployed for finding minimal set covers, together with DC axioms for branch pruning. Chu et al. [38] also extend FASTDC to discover approximate DCs (A-FASTDC) and constant DCs (C-FASTDC). Since FASTDC is sensitive to the number of records in the dataset, Pena and Almeida present BFASTDC [136], a bitwise version of FASTDC that uses logical operations to form the auxiliary data structures from which DCs are mined. Bleifuß et al. [21] propose a new algorithm Hydra, which overcomes the quadratic runtime complexity in the number of tuples in a relation. Based on the FASTDC algorithm, a system for discovering DC rules is implemented, namely, RuleMiner [40].

Application

Denial constraints are useful for detecting violations and enforcing the correct application semantics. Linear denial constraints [17, 18] are utilized to fix the numerical attributes in databases. Efficient approximation algorithms to obtain a database repair are presented [123]. Since DCs subsume existing formalisms and can express rules involving numerical values, with predicates such as "greater than" and "less than", [39] propose a holistic repairing algorithm under the constraints of DCs. It is worth noting that in practice, both the given DC rules and the data could be dirty. The V-repair model [110] proposes to change values with violations to other constants or to variables. Chu et al. [159] study the simultaneous repairing of both DCs and data, using the V-repair model as well.

4.5   Sequential Dependencies (SDs)

The above dependencies for order relationships, such as OFDs, ODs and DCs, only show the trends of data values. To show specific increase or decrease ranges, Golab et al. [87] propose sequential dependencies (SDs), which generalize ODs to express interesting relationships between ordered determinant attributes and distances on dependent attributes. Similar to MFDs and DDs, a distance metric is introduced in SDs.

Definition

A *sequential dependency* (SD) is in the form of

$$SD : X \rightarrow_g Y,$$

where $X \subseteq R$ is a set of ordered attributes, $Y \subseteq R$ can be measured by certain distance metrics, and g is an interval. It states that when tuples are sorted on X, the distances between the Y-values of any two consecutive tuples are within interval g.

Example

For the relation instance $r_{4.2}$ in Table 4.2, we can set an SD,

$$sd_1 : \text{nights} \rightarrow_{[100,200]} \text{subtotal}.$$

It identifies that subtotal raises within the interval [100, 200] with the increase of nights. As shown in Table 4.2, tuples are sorted on nights. Tuples t_2 and t_3 have "2" and "3" nights, respectively. The corresponding distance/increase of subtotal is $540 - 370 = 170$ within the range of [100, 200].

Special Case: ODs

As stated in Fig. 1.1, SDs subsume ODs in Sect. 4.2 as special cases. The interval g on distance in an SD can be used to express the order relationships, such as $[0, \infty)$ or $(-\infty, 0]$. Thereby, we rewrite od_1 in Sect. 4.2 as an SD,

$$sd_2 : \text{nights} \rightarrow_{(-\infty,0]} \text{unit-price}.$$

It means that unit-price for a room decreases with the increase of nights that a guest will stay. Therefore, SDs subsume the semantics of ODs, or SDs generalize/extend ODs, denoted by the arrow from ODs to SDs in Fig. 1.1.

Discovery

To discover reasonable SDs, Golab et al. [87] first define the confidence of an SD, i.e., the minimum number of tuples that needs to be removed or inserted to make the SD hold in a given dataset. It is worth noting that tuple insertion may apply to satisfy the distance constraints in an SD, different from the g_3 error measure of AFD in Sect. 2.5 with tuple deletion only to make an FD hold. Efficient computation of confidence is then studied for SD discovery, e.g., simple SDs of the form $X \rightarrow_{(0,\infty)} Y$ where Y always increases with X.

Application

By indicating the relevance of sequential attributes, SDs can discover missing data with large gaps, and extraneous data with small gaps and disorder data.

SDs are useful in network monitoring, e.g., auditing the polling frequency [87]. An SD : pollnum $\rightarrow_{[9,11]}$ time could be employed. It requires that the data collector probe the counters in about every 10 s. Too frequent polls, with time interval less than 9, or missing data, with time interval greater than 10, may indicate problems of the collector.

4.6   Conditional Sequential Dependencies (CSDs)

Real data sets, especially those with ordered attributes, are heterogeneous in nature. For the same attributes, constraints may not be appropriate for all tuples. Instead, they may fit for only one or several intervals. In this case, SDs cannot express the constraints specifically. Analogous to CFDs extending FDs in Sect. 2.7 and CDDs extending DDs in Sect. 3.4, in order to make SDs valid in a subset of tuples, SDs can also be extended with condition, i.e., only valid over a subset of tuples, namely conditional SDs [87]. CSDs declare SDs that conditionally hold in a period, considering that the frequency of data feed varies with time and the measurement attributes fluctuate with time. Different from the conditions that are specified by constants over categorical or heterogeneous data, it is natural to use a range to declare the conditions over ordered data.

Definition

A *conditional sequential dependency* (CSD) is a pair

$$\text{CSD} : X \rightarrow_g Y, t_r,$$

where t_r is a range pattern tuple and $X \rightarrow_g Y$ is an embedded SD, i.e., $X \subseteq R$ is a set of ordered attributes, and the distance of $Y \subseteq R$ between any two consecutive tuples should be within the interval g. Each range pattern t_r specifies a range of values of X, which can identify a subset of tuples over R (subsequence on X).

Example

For Table 4.2, consider a CSD as follows,

$$\text{csd}_1 : \text{nights} \rightarrow_{[30,35]} \text{taxes}, [2, 4].$$

It states that for two consecutive values of nights in the range of $[2, 4]$, their taxes value distance should be within $[30, 35]$. Intuitively, in a hotel, the taxes increase with the increase of nights. However, for promotion, the taxes should not increase too much, i.e., in a proper range of $[30, 35]$, if the customer stays for 2–4 nights. Consequently, tuple t_3 has "3" nights with "108" taxes, and t_4 has "4" nights with "140" taxes. Their distance on taxes is 32 within the range of $[30, 35]$.

Special Case: SDs

Figure 1.1 shows that CSDs subsume SDs as special cases. When there is no constraint specified in CSD conditions, i.e., $(-\infty, \infty)$, it is naturally an SD. For example, we can represent sd_1 in Sect. 4.5 as a CSD,

$$\text{csd}_2 : \text{nights} \rightarrow_{[100,200]} \text{subtotal}, (-\infty, \infty).$$

In this sense, CSDs subsume the semantics of SDs, or CSDs generalize/extend SDs, denoted by the arrow from SDs to CSDs in Fig. 1.1.

Discovery

The CSD tableau discovery framework discovers pattern tableaux for embedded SDs, which is used to specify subsets of data with conditions. It consists of two phases, (1) generating candidate intervals and (2) choosing from these candidates a small subset providing suitable (global) support. Therefore, similar to the support measure of CFDs in Sect. 2.7, evaluating how many tuples the constraint can cover, the support measure of CSDs is also studied [87]. For CSD tableau discovery, an exact dynamic programming algorithm for the tableau construction takes quadratic time in the number of candidate intervals. Instead of computing all possible intervals as candidates, efficient pruning and approximation algorithms are also developed for discovering CSDs. In particular, analogous to the distance ranges in DDs discovery in Sect. 3.3 or MDs discovery in Sect. 3.9, to avoid redundant semantics,

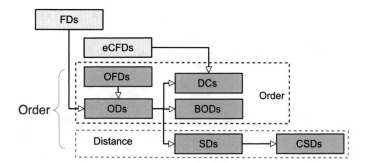

Fig. 4.1 A family tree of extensions between ordered data dependencies

the range pattern tableau of CSDs is expected to be minimized, but still with sufficient confidence and support. Approximation algorithms are devised for efficient CSD discovery. An approximation tolerance parameter δ is used for generating intervals. When the embedded SDs are supplied, the complexity of the discovery is $O((n \log^2 n)/\delta)$, in terms that n is the size of a relation and $1 + \delta$ is a bound on the approximation error.

Application

In addition to the applications of SDs in Sect. 4.5, CSDs with more expressive power have more application scenarios [87]. For instance, counters may reset if the router is rebooted. It leads to violations of the SD with the increase of time continually. A range pattern tableau is specified to denote the time intervals between reboots within which the SDs hold. In this sense, CSDs are useful in consistently summarizing and detecting problems. For instance, if there are too many short-range patterns, it may suggest premature counter roll-over.

4.7   Summary and Discussion

For ordered data, data dependencies are extended with operators $<, \leq, >, \geq$ for expressing the order relationships between data values. The ideas of conditional constraints for categorical data and distance metrics for heterogeneous data are also employed for data dependencies over ordered data, such as CSDs with conditions on determinant attributes and distances on dependent attributes as introduced in Sect. 4.6. Referring to Table 4.1 and Fig. 4.1, we summarize the relationships among the extensions over ordered data in different (sub)categories as follows.

Order

While OFDs [134, 135] specify that values in attributes should be ordered similarly, e.g., the mileage increases with time, ODs [50, 83–85] are introduced to express the constraints with different orders, such as the price of production drops with the increase of time. Rather

OFDs and ODs which only use $<, \leq, >, \geq$ to express orders without specifying concrete ranges, DCs [17, 18] study a more general form with $\{=, \neq, <, >, \leq, \geq\}$. The general form studied in [38] can further capture the semantics of other data dependencies such as FDs and CFDs.

Distance

In addition to the ordering relationships, the specific differences between ordered values are also promising to study. To this end, similar to DDs specifying constraints on distances between ordered values, SDs [87] study both the order of numerical data (e.g., timestamps) as well as the distance. Moreover, following the same line of CFDs extending FDs with conditions, CSDs [87] also study specified value range of the determinant X as a condition of SDs, so as to extend SDs for a subset of the whole relation.

Temporal Data

5

Temporal data, varying over time [101], are prevalent in the industry. For example, all kinds of sensor devices capture data from the physical world uninterruptedly. However, the sensor devices are often unreliable and produce missing and unreliable readings [100]. As a result, temporal data are often large and dirty. Temporal data quality issues are unique in challenges due to the presence of autocorrelations, trends, seasonality, gaps [46], and expected to be explained [93, 122]. To solve the problems of data management and data quality, various types of data dependencies on temporal data are proposed.

In this chapter, we introduce several typical works on the extensions over temporal data. The relationship of these data dependencies is shown in Fig. 1.1. Again, each pair of a data dependency and its special case, e.g., TDs extend ODs, corresponds to an arrow from ODs to TDs in the family tree.

Table 5.1 further categorizes these temporal constraints into the following sub-categories. (1) Relational: To adapt data dependencies to temporal data, a natural idea is to restrict them in the dimension of time, analogous to CFDs conditioning FDs in a subset of tuples. In other words, they still have a form of X determining Y with time functions or conditions. (2) Time series: Instead of timestamps serving as conditions, the values may be directly determined by time. Unlike the sequential dependencies that only consider values with order and value changes, the restrictions of speed on value changes within a period of time are also studied, i.e., on both time and value changes. (3) Event: Finally, while most data dependencies above consider only the values of each tuple, these data dependencies may ignore the constraints between timestamps themselves. To this end, temporal constraints are declared over the timestamps of events.

© The Author(s), under exclusive license to Springer Nature Switzerland AG 2023   85
S. Song and L. Chen, *Integrity Constraints on Rich Data Types*,
Synthesis Lectures on Data Management,
https://doi.org/10.1007/978-3-031-27177-9_5

Table 5.1 Category and special cases of data dependencies over temporal data, where each pair corresponds to an arrow in Fig. 1.1 such as SCs◁–ACs

Subcategory	Sections	Data dependencies	Special cases
Relational	5.1	TFDs	FDs
Relational	5.2	TDs	ODs
Timeseries	5.3	SCs	
Timeseries	5.4	MSCs	SCs
Timeseries	5.5	ACs	
Event	5.6	TCs	
Event	5.7	PNs	

5.1 Temporal Functional Dependencies (TFDs)

In the real world, there are errors that can be identified only through temporal constraints, e.g., FDs that restrict the rule on the temporal dimension. For example, a domain expert may come up with a rule stating that a person cannot be reported as arriving in two countries at the same time. To this end, Abedjan et al. [1], Jensen et al. [102], Wijsen et al. [178] study temporal functional dependencies. It can restrict that an exact entity cannot have two different events at the same time.

Definition

Let T be a time attribute in relation scheme R. A *temporal functional dependency* (TFD) over R is an expression

$$\text{TFD} : X \wedge \Delta \rightarrow Y,$$

where Δ is a time interval with a pair of minimum and maximum value, specified over the time attribute T. It states that for all pairs of tuples $t_i, t_j \in R$, s.t. $t_j[T] - t_i[T] \in \Delta$, the FD $X \rightarrow Y$ holds, i.e., $t_i[X] = t_j[X]$ implies $t_i[Y] = t_j[Y]$. A TFD means that during a time interval Δ, two tuples with the same X must have the same Y.

Example

In Table 5.2, a temporal functional dependency in the relation hotel can be express as

$$\text{tfd}_1 : \text{name} \wedge (0, 1 \text{ year}) \rightarrow \text{star},$$

which describes the rule "a hotel cannot be released with two different stars in a year". As shown in Table 5.2, we can find tuple t_3 and t_6 violate this dependency, since in June 2020 the star of "NC" is "3" and in December 2020 the star of "NC" is "4".

Table 5.2 Relation instance $r_{5.2}$ of hotel

	Time (year)	Name	Star	Price
t_1	2019.6	St. Regis	3	300
t_2	2020.6	NC	3	300
t_3	2020.6	NC	3	299
t_4	2020.6	Christina	4	399
t_5	2020.6	WD	4	500
t_6	2020.12	NC	4	399

Special Case: FDs

As mentioned in Fig. 1.1, TFDs extend FDs. The fd_1 in Sect. 1.1 can be represented as follows,

$$tfd_2 : address \wedge (0, \infty) \rightarrow region.$$

If there is a time attribute in Table 1.1, $(0, \infty)$ means that in any time, an address cannot correspond to two different regions. In this sense, TFDs subsume FDs, in other words, TFDs generalize/extend FDs, denoted by the arrow from FDs to TFDs in Fig. 1.1.

Axiomatization

For the class of temporal data, Wijsen [176] provides a complete and sound axiomatization of logical implication for TDs. There are seven inference rules for TDs. Let U be a set of attributes and Ψ, Φ, Υ be directed attribute sets (DAS) with $[\Psi], [\Phi], [\Upsilon] \subseteq U$. Let $A \in U$.

> TD1 (Augmentation) : If $\Phi \rightarrow_\alpha \Psi$, then $\Phi \sqcup \Upsilon \rightarrow_\alpha \Psi \sqcup \Upsilon$.
>
> TD2 (Transitivity) : If $\Phi \rightarrow_\alpha \Psi$ and $\Psi \rightarrow_\beta \Upsilon$, then $\Phi \rightarrow_{\alpha \cap \beta} \Upsilon$.
>
> TD3 (Upward heredity) : If $\Phi \rightarrow_\alpha \Psi$ and $\Phi \rightarrow_\beta \Psi$, then $\Phi \rightarrow_{\alpha \cup \beta} \Psi$.
>
> TD4 (Emptiness) : $\Phi \rightarrow_{\{\}} \Psi$.
>
> TD5 (Simplification) : If $\Phi \sqcup (A, \theta) \rightarrow_\alpha (A, \eta)$, then $\Phi \rightarrow_\alpha (A, \overline{\theta} \cup \eta)$.
>
> TD6 (Reciprocity) : If $\Phi \rightarrow_\alpha \Psi$, then $\widehat{\Phi} \rightarrow_{\alpha \cap Current} \widehat{\Psi}$.
>
> TD7 (False premise) : $(A, \perp) \rightarrow_{Future} \Phi$.

The first two rules (Augmentation and Transitivity) extend Armstrong's axioms. It is proved that none of the axioms of the axiomatization is redundant and additional nine useful inference rules are derived in [176].

Discovery

Abedjan et al. [1] propose a method for discovering valid TFDs in instances. Since Web data is inherently noisy, they are interested in the approximation of the problem, that is, all valid TFDs are detected, if the support value of the rule is higher than a given threshold, the rule is valid. For this problem, they provide a system AETAS to discover TFDs in Web data. Given a noisy dataset, they discover approximate functional dependencies firstly, i.e., traditional FDs maintaining most of the relationship for a given atomic duration. Using atomic duration can remove the temporal aspect of the relation, therefore, dependencies can be discovered purely based on record attributes. A set of AFDs can be ranked according to their level of support to help users verify. Users can reject the proposed AFD or verify it as a simple FD or TFD. For validated TFDs, they will discover the corresponding time interval, including values that apply only to specific entities. Because the data is dirty, they cannot just check the continuous events of each entity and collect the minimum duration. Therefore, they calculate the distribution of durations and mine it to determine the minimum duration that will eventually cut-off outlying values (i.e., invalid data). This minimum duration is then assigned to the AFD definition δ.

Application

Temporal functional dependencies can be used to detect errors in the relations consisting of event type attributes and time attributes. Given the TFD rule, the system AETAS in [1] mines the duration that leads to identifying temporal outliers. The problem of the sparseness for the data with value imputation can be tackled by the system. Moreover, the system can enforce the rule in the smallest meaningful time bucket to reduce the noise.

In addition, the PRAWN integration system [4] can use TFD to detect errors, and the record link system for temporal data can mine temporal behavior with repair-based duration discovery. In fact, the goal is to identify records that describe the same entity over time, and it is crucial to understand how long a value should hold.

5.2   Trend Dependencies (TDs)

Wijsen [176, 177] extends order dependencies with a time dimension for temporal databases. Rather than TFDs expressing the attributes only with $=$, i.e., TFDs only specify the attributes in two tuples to be the same with a time interval, a trend dependency allows attributes with linearly ordered domains to be compared over time by using any operator of $\{<, =, >, \leq, \geq, \neq\}$.

Definition

A *trend dependency* (TD) over R is a statement

$$\text{TD} : \phi[X] \rightarrow_\alpha \phi[Y],$$

where ϕ is the function with operator of $\{<, =, >, \leq, \geq, \neq\}$, α is a time function in time attribute, including $\text{Current} = \{(i, i)|i, i \in \mathbb{N}\}$, $\text{Future} = \{(i, j)|i, j \in \mathbb{N} \text{ and } i \leq j\}$ and $\text{Next} = \{(i, i + 1)|i \in \mathbb{N}\}$. The time attribute is represented by a set of natural numbers $\mathbb{N}(= \{1, 2, 3, \dots\})$. The time function α can be used to express the meaning that the next time point of i is $i + 1$. A TD states that, for any two tuples, when their attributes in X satisfy function $\phi[X]$ and their timestamps satisfy the time function α, then their attributes in Y must satisfy the function $\phi[Y]$.

Example

As shown in Table 5.3, a TD states that

$$td_1 : (SS\#, =) \rightarrow_{Next} (Sal, \leq).$$

It means that the salaries of employees should never decrease in consecutive time points. We can compare employee records at time i with records at the next time $i + 1$, for each time point i. Table 5.3 satisfies td_1, since all tuples satisfy this constraint. For instance, for t_1 and t_6 with same SS# of "A1", t_1 at time "1" has the Sal "100", while t_6 at time "2" has the Sal "110", i.e., increasing.

Similar to td_1, the following

$$td_2 : (SS\#, =) \rightarrow_{(Rank, \leq)}$$

also means that the ranks of employees should never decrease. The difference between them is that td_2 can compare the employee records at time i with records at any other future time $j, i < j$. For example, in Table 5.3, t_1 and t_{11} satisfy td_2. Specifically, with the same SS# of "A1", the Rank is "1" at time "1", and the Rank is "3" at time "3".

Table 5.3 Relation instance $r_{5.3}$ of employee

	Time	SS#	Rank	Sal
t_1	1	A1	1	100
t_2	1	B2	1	120
t_3	1	C3	3	140
t_4	1	D4	2	80
t_5	1	E5	2	120
t_6	2	A1	2	110
t_7	2	B2	2	110
t_8	2	C3	2	130
t_9	2	D4	3	90
t_{10}	2	E5	3	120
t_{11}	3	A1	3	120
t_{12}	3	B2	3	120

Special Case: ODs

In Fig. 1.1, we can find that TDs generalize ODs in Sect. 4.2. When there is only one version
of time, it is exactly an OD. The example od_1 shown in Sect. 4.2, can be transformed into
a TD as follows:

$$td_3 : (\text{nights}, \geq) \rightarrow_{\text{Current}} (\text{unit-price}, \leq).$$

Because there is no time attribute in Table 4.2, we assume that all the tuples are at the same
time i. Then the function *Current* can be employed to express the meaning of "at the same
time". The td_3 states that less nights means higher unit-price at the same time. Thus, TDs
subsume the semantics of ODs, or TDs generalize/extend ODs, denoted by the arrow from
ODs to TDs in Fig. 1.1.

Discovery

Wijsen [177] studies the problem TDMINE for the purpose of TD mining. Given a temporal
database, it mines the TDs that satisfy a given template and whose support and confidence
exceed certain threshold values. The time attribute function α can be given by finite enu-
meration. There is a decision problem TDMINE(D) to ask whether a specified support and
confidence can be attained by some TDs of a given TD class. Obviously, TDMINE is at least
as hard as TDMINE(D). TDMINE(D) can be solved in a brute force manner by an exhaus-
tive algorithm that computes the support and the confidence of each TD. The complexity
of TDMINE is studied, as well as algorithms to solve the problem. TDMINE can be solved
in polynomial time when time requirements are expressed as a function. Although solving
TDMINE is generally expensive, an interesting variant with acceptable time requirements
is worked out.

Application

TDs can express significant temporal trends. They can capture a significant family of data
evolution regularities, for example, salaries of employees generally do not decrease with the
passage of time. TDs enable attributes with linearly ordered domains to be compared over
time by comparison operator. With temporal semantics of *Current, Future, Next*, TDs can
express other types of data dependencies with time or order proposed in the literature, such
as ODs.

5.3   Speed Constraints (SCs)

Song et al. [158] study stream data constraints and propose *speed constraints* for stream
data cleaning. Unlike the sequential dependencies that only consider values with order and
value changes, speed constraints consider the restrictions of speed on value changes within
a period of time, i.e., both time and value changes. The rationale behind is that the change of

values is often constrained in a period of time, e.g., the maximum walking speed of a person, daily limit in financial and commodity markets, temperatures in a week, fuel consumption, etc.

Definition

A *speed constraint* (SC) can be defined as

$$SC : T \rightarrow_s A,$$

where T is a time attribute and A is a value attribute, i.e., each timestamp in T corresponds to a value in A. And $s = (s_{min}, s_{max})$ is a tuple of the minimum speed s_{min} and the maximum speed s_{max} on the relation R with window size w, where $T, A \in R$. Let $t_i[A]$ be the value of the i-th tuple, with a timestamp $t_i[T]$. Any two tuples $t_i[A], t_j[A]$ in a window, i.e., $0 < t_j[T] - t_i[T] \le w$, satisfy the speed constraint s, if they have

$$s_{min} \le \frac{t_j[A] - t_i[A]}{t_j[T] - t_i[T]} \le s_{max}.$$

A reasonable window size w is often expected, e.g., to consider the maximum walking speed in hours, rather than in years, since a person usually cannot keep on walking for several years.

Example

Given a relation in Table 5.4, a speed constraint can be

$$sc_1 : \text{time} \rightarrow_s \text{value},$$

where $s = (-0.5, 0.5)$, $w = 2$. Tuples t_3 and t_4 are in the window, and the speed of value change in tuples t_3 and t_4 is

$$\frac{t_4[\text{value}] - t_3[\text{value}]}{t_4[\text{time}] - t_3[\text{time}]} = \frac{10 - 13}{5 - 3} = -1.5 < -0.5.$$

Table 5.4 Relation instance $r_{5.4}$ of sequence

	Time	Value
t_1	1	12
t_2	2	12.5
t_3	3	13
t_4	5	10
t_5	7	15
t_6	8	15.5

Fig. 5.1 An example of speed
constraints

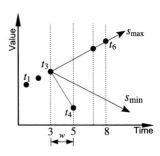

Thereby, they are identified as violations to $s = (-0.5, 0.5)$. As shown in Fig. 5.1, which plots the tuples in Table 5.4, the x-axis denotes Time, and the y-axis is Value. The window size $w = 2$ is also shown. Each tuple corresponds to a black point. We can easily find that the tuples t_3 and t_4 violate the speed constraints, denoted by arrows, since t_4 is out of the range constituted by the two arrows of s_{max} and s_{min}.

Discovery

Except for the natural speed constraints provided by users who know the dataset well, Song et al. [158] propose the method for discovering speed constraints in a particular domain where speed knowledge is not available. The statistical distribution of speeds is considered by sampling data pairs over datasets. In practice, confidence intervals are typically stated at the 95% or a higher confidence level, i.e., 95% or more of the speeds are regarded as accurate. Alternatively, a method with the principle of 3-sigma can be used to choose the speed constraints to keep 99.73% accurate data.

Application

Owing to unreliable sensor reading, or erroneous extraction of stock prices, the stream data are often dirty. Based on SCs, Song et al. [158] propose a linear time and constant space cleaning approach for stream data cleaning, toward local optimum under an efficient *Median Principle*. The algorithm can avoid changing those originally correct/clean data, i.e., observe the *minimum change principle* in data cleaning. To be specific, this algorithm can use speed constraints to attach candidates for each point that would be repaired. Then the median principle is used to choose the repair with the minimum cost. In addition to the repairing application, the speed constraints can also be used to detect errors, i.e., the points violating speed constraints can be regarded as errors.

5.4   Multi-speed Constraints (MSCs)

While speed constraints (SCs) are restricted to a single range between the maximum and the minimum, multi-speed constraints [82] have multiple intervals of speed, making the constraints more specific and reasonable. The constraint result is defined for a given window and can be flexibly extended to the whole time series in actual application scenarios. It also makes the repair method more widely applicable.

Definition

A *multi-speed constraint* (MSC) can be defined as

$$\text{MSC} : T \rightarrow_{ms} A,$$

where T is a time attribute and A is a value attribute, i.e., each timestamp in T corresponds to a value in A. The multi-speed constraint $ms = \{s_1, s_2, \ldots, s_n\}$ with a given time window w refers to a set of constraint intervals s_r ($r = 1, 2, \ldots, n$), where $s_r = \{s_r^{\min} s_r^{\max}\}$ is a pair of $s_r^{\min}$ and $s_r^{\max}$.

Let $t_i[A]$ be the value of the i-th tuple, with a timestamp $t_i[T]$. Any two tuples $t_i[A]$, $t_j[A]$ in a window, i.e., $0 < t_j[T] - t_i[T] \le w$, satisfy the multi-speed constraint ms, if they have

$$s_r^{\min} \le \frac{t_j[A] - t_i[A]}{t_j[T] - t_i[T]} \le s_r^{\max},$$

for some $r = 1, 2, \ldots, n$.

Example

Given a relation in Table 5.5, a multi-speed constraint can be

$$\text{msc}_1 : \text{time} \rightarrow_{ms} \text{value},$$

where $ms = \{s_1, s_2\}$, $s_1 = [-0.5, -0.2]$, $s_2 = [0.2, 0.5]$, $w = 5$. Tuples t_3, t_4, t_5, t_6, t_7 and t_8 are in the window. The speed of value change in tuples t_3 and t_4 is

$$\frac{t_4[\text{value}] - t_3[\text{value}]}{t_4[\text{time}] - t_3[\text{time}]} = 0.2 < \frac{13.3 - 13}{4 - 3} = 0.3 < 0.5.$$

Table 5.5 Relation instance $r_{5.5}$ of sequence

	Time	Value
t_1	1	12
t_2	2	12.5
t_3	3	13
t_4	4	13.3
t_5	5	10
t_6	6	13
t_7	7	12
t_8	8	15.5

Fig. 5.2 An example of
multi-speed constraints

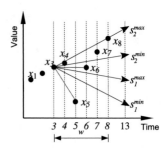

Thereby, they are not identified as violations to $s_2 = [0.2, 0.5]$, i.e., tuples t_3 and t_4 meet
the multi-speed constraints ms. And the same is true for t_3 and t_7. Nevertheless, the speed
of value change in tuples t_3 and t_6 is

$$\frac{t_6[\text{value}] - t_3[\text{value}]}{t_6[\text{time}] - t_3[\text{time}]} = -0.2 < \frac{13 - 13}{6 - 3} = 0 < 0.2.$$

Thereby, they are identified as violations to the multi-speed constraints ms, which is the
same as t_3 and t_5.

As shown in Fig. 5.2, which plots the tuples in Table 5.5, the x-axis denotes time, and the
y-axis is value. The window size $w = 5$ is also shown. Each tuple corresponds to a black
point. We can easily find that the tuples t_3 and t_5 and the tuples t_3 and t_6 violate the speed
constraints, denoted by arrows, since t_5 and t_6 are out of the range constituted by the two
pairs of arrows, i.e., $s_1^{\max}$ and $s_1^{\min}$, $s_2^{\max}$ and $s_2^{\min}$.

Special Case: SCs

As shown in Fig. 1.1, MSCs are extension of SCs. Indeed, multi-speed constraint with one
interval is equivalent to speed constraint. For the relation instance in Table 5.4, sc_1 in Sect. 5.3
can be equivalently represented by

$$msc_1 : time \rightarrow_{ms} value,$$

where $ms = \{s_1\}$, $s_1 = [-0.5, 0.5]$, $w = 5$.

Discovery

Gao et al. [82] present a MSC discovery method in a particular window. With the current point
x_k and its previous points in the same window, the method generates the speed constraint
ranges by the tuples. For each speed constraint range, we can refer to SCs and adopt the
method based on the 3-sigma principle to maintain a higher confidence level. Then the
multi-speed constraint ranges of x_k are combined into a speed constraint set.

Application

Gao et al. [82] introduce multi-speed constraints based on the speed of data change, and then the time series data can be constrained to detect the anomalies. Meanwhile, candidate repairing points for each data point in the whole window of time series can be generated by the multi-speed constraints with its subsequent points. The paper proposes a time series dynamic programming-based repairing method under multi-speed constraints. And the method under multi-speed constraints identifies the optimal repairing path according to dynamic programming in the given candidate points, which is equivalent to or better than SCREEN [158] in a special case of single-speed constraint.

5.5   Acceleration Constraints (ACs)

While speed constraints above can be used to detect and repair errors, some subtle data changes might not be detectable by speed constraints. For example, a vehicle may need 5 s to accelerate from 0 to 100 km/h. It may not exceed the maximum speed of 100 km/h. However, if the acceleration is observed in less than 5 s, some errors may occur. In this sense, the acceleration constraints [154] are also necessary in addition to the speed constraints.

Definition

Given a window size w, an *acceleration constraint* (AC) is

$$AC : T \rightarrow_a A,$$

where $a = (a_{min}, a_{max})$ is a pair of the minimum acceleration a_{min} and the maximum acceleration a_{max}, T is a time attribute and A is a value attribute over the relation R, $T, A \in R$. Let $t_i[A]$ be the value of the i-th tuple, with a timestamp $t_i[T]$, i.e., each timestamp from T corresponds to a value in A. If $0 < t_j[T] - t_i[T] < w$, i.e., for any $t_i[A], t_j[A]$ in a window,

and $a_{min} < \dfrac{\frac{t_j[A]-t_i[A]}{t_j[T]-t_i[T]} - \frac{t_i[A]-t_{i-1}[A]}{t_i[T]-t_{i-1}[T]}}{t_j[T]-t_i[T]} < a_{max}$, then t_i, t_j satisfy the acceleration constraint a.

Example

Consider another relation in Table 5.6, the acceleration constraint can be defined as

$$ac_1 : \text{Time} \rightarrow_a \text{Value},$$

where $w = 2$, $a = (-1, 1)$, i.e., $a_{min} = -1$, $a_{max} = 1$. Similarly, Fig. 5.3 plots the tuples in Table 5.6. The tuples are shown as the black points. The maximum acceleration constraint a_{max} is the red line, and the minimum acceleration constraint a_{min} is the blue line (since the acceleration involves three points, the acceleration lines are curve). We can find that this relation cannot satisfy ac_1 above, since the tuple t_4 is out of the acceleration constraint

Table 5.6 Relation instance $r_{5.6}$ of time series

	Time	Value
t_1	1	0
t_2	2	0.5
t_3	3	2
t_4	4	6.3
t_5	5	6
t_6	6	7
t_7	7	8

Fig. 5.3 An example of acceleration constraints

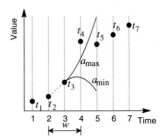

range. To be specific, with tuples t_2, t_3 and t_4, whose timestamps are within $t_4[T] - w$, the acceleration can be computed with the formula

$$\frac{\frac{t_4[\text{Value}] - t_3[\text{Value}]}{t_4[\text{Time}] - t_3[\text{Time}]} - \frac{t_3[\text{Value}] - t_2[\text{Value}]}{t_3[\text{Time}] - t_2[\text{Time}]}}{t_4[\text{Time}] - t_3[\text{Time}]} = \frac{\frac{6.3-2}{4-3} - \frac{2-0.5}{3-2}}{4 - 3} = 2.8.$$

It obviously violates the acceleration constraint $a = (-1, 1)$.

Discovery

Intuitively, the acceleration constraints can be found from the characteristics of data, for example, the acceleration of GPS data collected by a car cannot be too large. However, there are many other datasets that have no natural acceleration semantics. In this case, there should be a method to attach the acceleration constraints from the dataset values. To solve this problem, Song et al. [154] propose the acceleration constraints discovery method with the distribution of accelerations for each data point. Similar to the SC discovery, with the idea of most data to be correct as well as accelerations, the 3-sigma principle is also employed to select the acceleration constraints.

Application

To repair the data from GPS trajectories or some other sensors readings that violate acceleration constraints, Song et al. [158] propose an efficient *Median Principle* method with local optimum for stream data cleaning. To be specific, in order to repair the dirty data more accurately and repair the subtle error that speed constraints cannot be detected, this method applies both the acceleration constraints and speed constraints to get several repair candidates. Then it uses the Median Principle to find the local optimum from the candidates for each point in the data stream.

5.6 Temporal Constraints (TCs)

Instead of timestamps, most data dependencies like SDs in Sect. 4.5 or SCs in Sect. 5.3 above only consider the values of each tuple. These data dependencies may ignore the constraints between timestamps themselves. Dechter et al. [47] introduce *temporal constraint satisfaction problem* (TCSP). The temporal information is represented by a set of unary or binary constraints, each of which specifies a set of permitting intervals.

Definition

Given a relation scheme R with time attribute T, a unary *temporal constraint* (TC) can be represented by a set of intervals $\{[a_1, b_1], \ldots, [a_n, b_n]\}$ on timestamps as follows:

$$\text{TC}: \forall t_\alpha \in R, t_\alpha[A] = a \to (a_1 \leq t_\alpha[T] \leq b_1) \vee \cdots \vee (a_n \leq t_\alpha[T] \leq b_n),$$

where $A \in R$ and a is constant value of attribute A. It states that for any tuple t_α with value $t_\alpha[A] = a$, its timestamp $t_\alpha[T]$ should be either in the rage of $[a_1, b_1]$, ..., or $[a_n, b_n]$. In addition, a binary *temporal constraint* (TC) can be represented by a set of intervals $\{[a_1, b_1], \ldots, [a_n, b_n]\}$ on timestamp distances,

$$\text{TC}: \forall t_\alpha, t_\beta \in R, t_\alpha[A] = a \wedge t_\beta[A] = b \to$$
$$(a_1 \leq t_\alpha[T] - t_\beta[T] \leq b_1) \vee \cdots \vee (a_n \leq t_\alpha[T] - t_\beta[T] \leq b_n),$$

where $A \in R$ and a, b are constant values of attribute A. It states that for any two tuples t_α and t_β with A values $t_\alpha[A] = a$ and $t_\beta[A] = b$, their distance on timestamps $t_\alpha[T] - t_\beta[T]$ should be either in the rage of $[a_1, b_1]$, ..., or $[a_n, b_n]$.

Example

It is worth noting that the unary and binary temporal constraints form a temporal constraint network, where each value in attribute A corresponds to a node. Figure 5.4 presents an example temporal constraint network from [144], declared over the relation in Table 5.7. It denotes the steps (a.k.a. events, denoted by nodes) that are required in every part design process of a train manufacturer. The temporal constraint network (abstracted from workflow specifications) specifies the constraints on occurring timestamps of events.

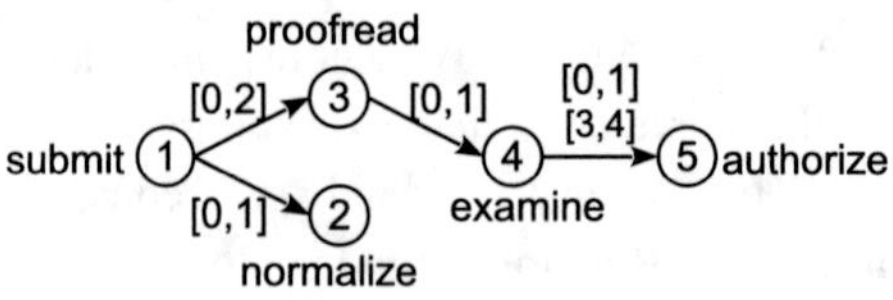

Fig. 5.4 An example temporal constraint network

Table 5.7 Relation instance $r_{5.7}$ for sequence data

	ID	Event	Time	Value
t_1	1	Submit	1	12
t_2	2	Normalize	2	12.5
t_3	3	Proofread	3	13
t_4	4	Examine	5	10
t_5	5	Authorize	13	15

Note that multiple intervals could be declared between two events. For instance, [0, 1], [3, 4] on edge $4 \rightarrow 5$ denote tc_1 as follows,

$$tc_1 : \forall t_\alpha, t_\beta \in R, t_\beta[\text{event}] = \text{"examine"} \wedge t_\alpha[\text{event}] = \text{"authorize"} \rightarrow$$
$$(0 \le t_\alpha[\text{time}] - t_\beta[\text{time}] \le 1) \vee (3 \le t_\alpha[\text{time}] - t_\beta[\text{time}] \ge 4).$$

It denotes that a tuple with event authorize can be processed after the tuple with event examine either by the department head in 0–1 h or by the division head in 3–4 h. For instance, consider the example instance of event trace in Table 5.7. The instance records the five steps (events) for processing a part design work, i.e., submit, normalize, proofread, etc. Each event has a timestamp, indicating the time when this event occurred. The tuples t_4 and t_5 violate this constraint, since $t_5[\text{time}] - t_4[\text{time}] = 13 - 5 = 8$ is out the range of [0, 1] or [3, 4].

Application

Dechter et al. [47] utilize temporal constraints to perform reasoning tasks as follows. It finds all feasible times that a given event can occur. Moreover, we may also find all the possible relationships between two given events. Finally, one can generate one or more scenarios consistent with the information provided. Likewise, serving as integrity constraints, temporal constraints are employed to repair the erroneous timestamps [144].

5.7   Petri Nets (PNs)

While TCs in Sect. 5.6 specify the constraints on timestamp, they do enforce whether the event should appear or not. To this end, Petri nets are employed for conformance checking van der Aalst [164]. Indeed, Petri net is often utilized as a mathematical modeling tool to describe the information processing systems. Moreover, Petri net is also used as a graphical tool to visualize the flow charts and networks of a business process. Finally, Petri net has also been recently considered as constraints in repairing event errors [94, 169].

Definition

Following the convention of notation in [164], a *Petri net* is in a form of a triplet $N(\mathcal{P}, \mathcal{T}, \mathcal{F})$, where $\mathcal{P}$ is a finite set of places, $\mathcal{T}$ is a finite set of transitions, and $\mathcal{F} \subseteq (\mathcal{P} \times \mathcal{T}) \cup (\mathcal{T} \times \mathcal{P})$ is a set of directed arcs (flow relation). For any node $t_i \in \mathcal{P} \times \mathcal{T}$, let $\bullet t_i = \{t_j | (t_j, t_i) \in \mathcal{F}\}$ denote the pre-set of t_i and $t_i \bullet = \{t_j | (t_i, t_j) \in \mathcal{F}\}$ be the post-set of t_i. The pre/post-set representation can be nested, e.g., $\bullet(\bullet t_i)$ denotes $\cup_{t_j \in \bullet t_i} \bullet t_j$, the union of the pre-sets of a pre-set.

A process specification is a Petri net $N_s(\mathcal{P}_s, \mathcal{T}_s, \mathcal{F}_s)$. There is a unique source place $\mathsf{b}_{start} \in \mathcal{P}_s$, $\bullet \mathsf{b}_{start} = \emptyset$, and a unique sink place $\mathsf{b}_{end} \in \mathcal{P}_s$, $\mathsf{b}_{end} \bullet = \emptyset$, whose post-set is empty. Each node $t \in \mathcal{P}_s \cup \mathcal{T}_s$, place or transition, is on a path from b_{start} to b_{end}. Each transition $e \in \mathcal{T}_s$ corresponds to an event in the execution of the process. An *event sequence* σ, denoted by a relation, with respect to a process specification $N_s(\mathcal{P}_s, \mathcal{T}_s, \mathcal{F}_s)$ is a finite sequence of events (transitions). Each sequence logs an execution of the process defined by N.

Again, following the notations in [170, 171], a firing sequence of process specification $N_s(\mathcal{P}_s, \mathcal{T}_s, \mathcal{F}_s)$ and its post-set, are defined recursively as follows: (1) An empty sequence ε is a firing sequence, and $\varepsilon \bullet = \{\mathsf{b}_{start}\}$; (2) If σ is a firing sequence, $e \in \mathcal{T}_s$ is a transition (event), and $\bullet e \subseteq \sigma$, then σe is also a firing sequence, and $(\sigma e)\bullet = (\sigma \bullet) - (\bullet e) + (e\bullet)$. We say that a sequence σ conforms to a process specification, denoted by $\sigma \vDash N_s$, if σ is a firing sequence w.r.t. N_s and $\sigma \bullet = \{\mathsf{b}_{end}\}$.

Besides logical controls specified by Petri nets, additional constraints could be further declared to restrict the occurrence of events. A *time constraint* of two consecutive events $e_i, e_j, e_i \in \bullet(\bullet e_j)$, denoted as $\delta(e_i, e_j)$, is the maximum distance of occurrence time of the event e_j and the most recent e_i that appear in any firing sequence. It states that

$$t_j[T] - t_i[T] \leq \delta(e_i, e_j),$$

where $t_i[T]$ is the occurrence time of event e_i, and $t_j[T]$ is the occurrence time of event e_j.

Example

Figure 5.5 presents an example of process specification, which corresponds to an engineering drawing process in a train manufacturer. Each square is a transition, while each circle denotes

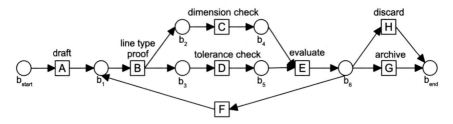

Fig. 5.5 An example constraint of Petri net

Table 5.8 Relation instance $r_{5.8}$ of event sequence

	Event	Value	Tme
t_1	A	Draft	1
t_2	B	Line type proof	2
t_3	C	Dimension check	3
t_4	D	Tolerance check	5
t_5	E	Evaluate	7
t_6	G	Archive	10

a place. The arrows attached to a transition denote that the corresponding flows should be executed in parallel.

Consider the process specification N_s in Fig. 5.5 and a sequence $\sigma =<$ ABCDEG $>$ ordered by time in Table 5.8. To investigate whether σ is a firing sequence with respect to N_s, we start from the empty sequence ε with $\varepsilon \bullet = \{b_{start}\}$. Since the first event A has $\bullet A = \{b_{start}\} \subseteq \varepsilon \bullet$, the augmentation $< A >$ is also a firing sequence with $< A > \bullet = \{b_1\}$. It follows $< AB > \bullet = \{b_2, b_3\}$. For the next C, as $\bullet C = \{b_2\} \subset < AB > \bullet$, the firing sequence becomes $< AB >$ with post-set $\{b_3, b_4\}$. Similarly, we have $< ABCD > \bullet = \{b_4, b_5\}$ by appending D. As $\bullet E = \{b_4, b_5\}$ and $E \bullet = \{b_6\}$, it leads to $< ABCDE > \bullet = \{b_6\}$, and finally $< ABCDEG > \bullet = \{b_{end}\}$. Therefore, the sequence $< ABCDEG >$ conforms to the specification.

A firing sequence can be verified whether conforms to the process specification with time constraints. For instance, consider a time constraint $\delta(A, B) = 2$ between consecutive events A and B in Fig. 5.5. Suppose that the occurrence time of event B is $t_2[time]$ in the sequence $\sigma =<$ ABCDEG $>$ in Table 5.8. For the most recent occurrence of event A in the sequence, the time constraint requires

$$t_2[time] - t_1[time] \le \delta(A, B).$$

As shown in Table 5.8, the occurrence times of events A and B is "1" and "2", respectively. It satisfies the constraint with $2 - 1 = 1 < \delta(A, B) = 2$.

Application

Traditionally, Petri nets are employed to specify the processes, such as personnel management processes, biological information, workflows and so on. Recently, using Petri nets as constraints, Wang et al. [170, 171] study the missing event recovery problem. The minimum recovery problem to find and impute missing events that can satisfy the constraints specified Petri net.

5.8   Summary and Discussion

Temporal data are characterized by data elements with time-varying information [101]. It is important in various scenarios such as IoT, where sensor devices uninterruptedly obtain data from the physical world. The sensor devices are often unreliable with missing readings [100] or even wrong timestamps [144]. As a result, time series data are usually very large, incomplete [170, 171] and dirty [169, 185]. While data dependencies over numerical data such as sequential dependencies [87] could be employed to partially capture the order information, more advanced constraints on temporal features are expected, e.g., to capture unique data quality challenges due to the presence of autocorrelations, trends, seasonality and gaps in the time series data [46]. Referring to Table 5.1 and Fig. 5.6, we summarize the relationships among the extensions over temporal data in different (sub)categories as follows.

Relational

To adapt data dependencies to temporal data, TFDs [1, 102] naturally restrict them in the dimension of time, analogous to CFDs conditioning FDs in a subset of tuples. In other words, they still have a form of X determining Y with time functions or conditions. In contrast, TDs [176, 177] extend ODs with a time dimension for temporal databases. Rather than TFDs expressing the attributes only with $=$, i.e., TFDs only specify the attributes in two tuples to be the same with a time interval, TDs allow attributes with linearly ordered domains to be compared over time by using any operator of $\{<, =, >, \leq, \geq, \neq\}$.

Fig. 5.6 A family tree of extensions between temporal data dependencies

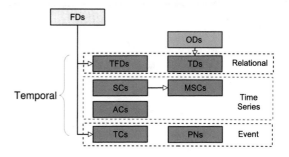

Timeseries

Unlike TFDs or TDs specifying timestamps as conditions, SCs [158] consider the values directly determined by time. Unlike SDs that only consider values with order and value changes, SCs restrict the speed of value changes within a period of time, i.e., on both time and value changes. Moreover, ACs [154] further consider the constraints on the acceleration of value changes with respect to time, in addition to SCs. The acceleration constraints are useful to detect the subtle data changes which are ignored by speed constraints.

Event

While most data dependencies above consider only the values of each tuple, TCs [47] further consider the constraints between timestamps of events. Temporal constraints are represented by a set of unary and binary constraints, which naturally form a temporal constraint network. In this sense, Petri net with more rich semantics of event occurrence is also employed for conformance checking of event sequences.

Graph Data

6

Graph data have been widely observed in real-world applications, e.g., knowledge bases and social networks can be modeled as graphs. In such scenarios, entities are represented by the vertexes in the graphs, each of which has one class tag such as persons, attribute values and connections with other entities. With the consideration over three main components, class, relation and attribute, metalanguage for graph models is introduced to define how graph data is serialized and compiled in files or databases, e.g., eXtensible Markup Language (XML) and Resource Description Frameworks (RDFs). The RDF view of graphs treats graph data as a collection of triples with the form ⟨subject, relation, object⟩, where each vertex shown in subject, object belongs to one class, and the attributes of each vertex are represented by RDF triples. In general, graph data can be represented as a tuple $G = (V, E, L)$ with V, E, L denoting the set of vertexes, edges and labels, respectively, and is convertible with the RDF view. Unlike relational data, graph data do not have a specific schema. In this sense, novel types of integrity constraints need to be developed for graph data.

The graph part in Fig. 1.1 presents the major data dependencies extended for graph data. The corresponding special cases, where the data dependencies are extended from, are presented as well. Each pair of a data dependency and its special case, e.g., XCFDs extend XFDs, corresponds to an arrow from XFDs to XCFDs in the family tree in Fig. 1.1.

Table 6.1 further categorizes these constraints on graph data into the following subcategories. (1) Path: To specify constraints on graphs, a natural idea is to consider the neighborhood of vertexes, in particular, whether the labels can appear in adjacent vertexes. Such a constraint on edges of two vertexes can be further extended to a path. (2) Tree: For the data in a tree structure, such as XML, the constraints can be declared over the (sub)trees, in addition to paths. Similar to functional dependencies, some attribute values may determine others in the tree. (3) Pattern: Finally, the constraints are extended from simple paths and trees to more complicated patterns. Again, (temporal) functional dependencies or even denial constraints can be declared over the subgraphs identified by the patterns.

© The Author(s), under exclusive license to Springer Nature Switzerland AG 2023
S. Song and L. Chen, *Integrity Constraints on Rich Data Types*,
Synthesis Lectures on Data Management,
https://doi.org/10.1007/978-3-031-27177-9_6

Table 6.1 Category and special cases of data dependencies over graph data, where each pair corresponds to an arrow in Fig. 1.1 such as GFDs◁–GEDs

Subcategory	Section	Data dependencies	Special cases
Path	6.1	NCs	DDs
Path	6.2	NLCs	NCs
Path	6.3	PLCs	NCs
Tree	6.4	XFDs	FDs
Tree	6.5	XCFDs	XFDs
Pattern	6.6	GKs	
Pattern	6.7	GPARs	
Pattern	6.8	GFDs	XCFDs
Pattern	6.9	GEDs	GKs, GFDs
Pattern	6.10	GARs	GEDs
Pattern	6.11	GDDs	GDDs
Pattern	6.12	GDCs	GEDs
Pattern	6.13	TGFDs	GFDs

6.1   Neighborhood Constraints (NCs)

A wide class of data, such as protein networks, workflow networks, can be modeled as graphs where vertex labels are data values. However, similar to other data values, vertex labels are often dirty owing to various reasons, such as typing errors or report errors of scientific experimental results. Moreover, the connections between vertexes could also be dirty, i.e., erroneous graph structures. In this sense, Song et al. [153] propose neighborhood constraints (NCs). By specifying the label pairs allowed to appear on adjacent vertexes, NCs can be applied to detect and repair erroneous vertex labels and graph structures.

Definition

A *instance graph* $G = (V, E, L)$ with a set of vertexes V and edges E has labels from L for all the vertexes in V, given a labeling function $\lambda : V \rightarrow L$, i.e., $\lambda(v) \in L, \forall v \in V$. Let $L = \{\ell_1, \dots, \ell_{|L|}\}$ denote the set of all labels.

A *neighborhood constraint graph* $S(L, N)$ is also a graph, where N specifies the pair-wise neighborhood constraints of unique labels in L. As stated in [153], two labels ℓ_1, ℓ_2 are said to match a constraint graph $S(L, N)$, denoted by $(\ell_1, \ell_2) \asymp S$, if they are either the same label, $\ell_1 = \ell_2$, or denote an edge in S, $(\ell_1, \ell_2) \in N$. In other words, we have $(\ell, \ell) \in N$ by default.

A *neighborhood constraint* (NC) is in the form of

$$NC : E \rightarrow N.$$

It states that each edge $(v, u) \in E$ should have labels $(\lambda(v), \lambda(u)) \in N$. Formally, a instance graph $G(V, E, L)$ *satisfies* a constraint graph $S(L, N)$, if $(v, u) \in E$ in the instance graph implies $(\lambda(v), \lambda(u)) \asymp S$ in the constraint graph, $\forall(v, u) \in E$. In other words, for any edge $(v, e) \in E$, their labels $\lambda(v), \lambda(u)$ must match the constraint graph S, i.e., having either $\lambda(v) = \lambda(u)$ or $(\lambda(v), \lambda(u)) \in N$.

Example

Figure 6.1a illustrates a constraint graph that specifies the neighborhood among four labels $\{a, b, c, d\}$, where each vertex denotes a unique label. Figure 6.1b illustrates an instance graph with four vertexes $V = \{1, 2, 3, 4\}$, where each vertex is associated with a label from $L = \{a, b, c, d\}$. In Fig. 6.1b, edge $(1, 3)$ with labels a, d indicates a violation to the NC

$$nc_1 : E \rightarrow N,$$

as a, d are not adjacent according to the constraint in Fig. 6.1a. Therefore, the instance graph cannot satisfy the constraint graph.

Special Case: DDs

From Fig. 1.1, we can find that there is an arrow from DDs to NCs, which means that NCs subsume DDs. Each DD can be represented as an NC because each attribute for tuples can be constructed as a graph. Take dd_1 in Sect. 3.3 as an example. There are edges between tuple with a distance of name equal to or less than 1, and similarly for attributes street or

Fig. 6.1 Neighborhood constraints and label repairing on similarity network

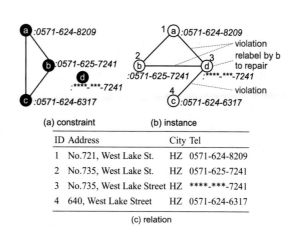

(a) constraint   (b) instance

ID	Address	City	Tel
1	No.721, West Lake St.	HZ	0571-624-8209
2	No.735, West Lake St.	HZ	0571-625-7241
3	No.735, West Lake Street	HZ	****-***-7241
4	640, West Lake Street	HZ	0571-624-6317

(c) relation

address with a distance equal to or less than 5. Indeed, there is a DD declared over the relation in Fig. 6.1c,

$$dd_4 : \text{Address}(\leq 6), \text{City}(\leq 0) \rightarrow \text{Tel}(\leq 5),$$

which leads to the instance and constraint graphs. It states that when two tuples have same City and the Address within the distance of 6, then the distance of Tel should also be equal to or less than 5. For example, t_1 and t_2 have same City, the distance of Address between t_1 and t_2 is 2, then the distance of Tel for them is 5. However, the tuples t_2 and t_3 violate this constraint since the distance Address is 4 and the distance for Tel is 7. In the corresponding Fig. 6.1a, there is an edge between a and b, but no edge between b and d. Consequently, NCs generalize/extend DDs, denoted by the arrow from DDs to NCs.

Application

The neighborhood constraints can be used for data repairing by eliminating violations to the integrity constraints. In order to address the violations in the instance graph, Song et al. [153] propose a repairing strategy that modifies at least one value (label/neighbor) of the vertexes v, u. For instance, as illustrated in Fig. 6.1, repairing vertex 3 by label b eliminates the violations to vertexes 1, 2, 4. The relabeling problem is proved to be NP-hard, together with several approximate repairing algorithms including greedy heuristics, contraction method and a hybrid approach.

In addition, Song et al. [155] devise a cubic-time constant-factor graph repairing algorithm with both label and neighbor repairs (given degree-bounded instance graphs). To put together the beauty of violation elimination heuristics and termination, an approach AlterGC cooperates contraction and greedy relabeling. When both label and neighbor errors exist, the proposed Grepair algorithm shows both higher repair accuracy and better time performance.

6.2   Node Label Constraints (NLCs)

Broder et al. [28] study a reachability problem with node label constraints (NLCs). Compared to the NC which only shows the relationship between two neighbors, NLCs study the path with labels in vertexes. It is motivated by the problem of evaluating graph constraints in content-based pub/sub-systems. To solve the problems of publish/subscribe to social networks, they create one entity in the content-based index for each vertex in the graph.

Definition

Consider a directed graph $G = (V, E, L)$, together with a labeling function $\lambda : V \rightarrow L$, i.e., $\lambda(v) \in L, \forall v \in V$. Let P be a path in the graph with edges $P \subseteq E$, and Q be a set of labels declared as constraints having $Q \subseteq L$.

A *node label constraint* (NLC) can be defined as

$$\text{NLC} : P \rightarrow Q.$$

It states that for each vertex v in the path P, its label must be in Q, $\lambda(v) \in Q$.

Example

For example, in Fig. 6.2, three advertising networks are connected. In this system, the three publishers can access to the four available advertisers. To ensure a good quality of recommendation, each ad network can specify targeting attributes. If a young user visits a sports page from a world publisher, she/he will only receive ads from the advertiser Adv_1. The reason is that only Net_1 satisfies the constraint with label "Sports", Pub_1 satisfies the constraint with label "World", and Adv_1 fits the constraint with label "Young". It can be presented by an NLC as follows:

$$\text{nlc}_1 : P \rightarrow \{\text{World, Sports, Young}\}$$

A path P with vertexes (Pub_1, Net_1, Adv_1) is said to satisfy nlc_1 with the label set $Q = \{\text{World, Sports, Young}\}$. Specifically, there are labels "World, Sports, Young" associated to vertexes Pub_1, Net_1 and Adv_1, respectively.

Special Case: NCs

As presented in Fig. 1.1, NLCs subsume NCs in Sect. 6.1. For nc_1 in Sect. 6.1, we can write a set of NLCs on single-edge paths as follows,

$$\text{nlc}_2 : (v, u) \rightarrow \{\lambda(v), \lambda(u)\}.$$

Fig. 6.2 Sample Web advertising exchange

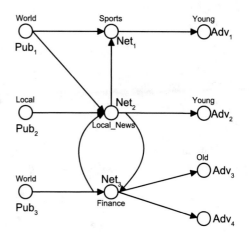

It states that the only edge $(v, u) \in E$ in the path P has vertexes with labels from a set $Q = \{\lambda(v), \lambda(u)\}$ where $(\lambda(v), \lambda(u)) \in N$. In this sense, NLCs subsume the semantics of NCs, or NLCs generalize/extend NCs, denoted by the arrow from NCs to NLCs in Fig. 1.1.

Application

There are connection graphs formed by the correlated users in social networks. For an application, each user subscribes and produces a series of interesting messages. Some network applications may be the vertexes for content dissemination. They can accumulate and redistribute information to the subscribers who are interested in the information. Broder et al. [28] use a method to solve the problem of node label constraints. The system can be presented by a directed graph, consisting of three types of vertexes: publishers, intermediaries and subscribers. To ensure that an event from a publisher can only be sent to subscribers on the same path, only the nodes without any incoming/outgoing edges can be publishers/subscribers.

6.3   Path Label Constraints (PLCs)

Both the NCs and NLCs only express the constraints on the labels of vertexes. However, in many other situations, there are labels on the edges. To this end, Jin et al. [103] propose path label constraints (PLCs), declaring constraints on edge labels. The label-constraint reachability query problem asks whether a vertex u can reach vertex v through a path whose edge labels are constrained by a set of labels.

Definition

Consider a labeled directed graph $G = (V, E, L)$, where V is the set of vertexes, E is the set of edges, L is the set of edge labels and λ is a function that assigns each edge a label, $\lambda : E \rightarrow L$, i.e., $\lambda(v, u) \in L, \forall(v, u) \in E$.

A *path label constraint* (PLC) can be expressed as

$$PLC : P \rightarrow Y,$$

where P is a path in graph G and $Y \subseteq L$ is a set of edge labels. It states that for each edge (v, u) in the path P, its edge label $\lambda(v, u)$ must belong to the set Y, i.e., $\lambda(v, u) \in Y$.

Example

For example, in Fig. 6.3, we can see a labeled directed graph $G = (V, E, L)$, with the set of vertexes V, edges E, and edge labels L. A PLC can be declared as follows:

$$plc_1 : P \rightarrow \{a, b, c\},$$

where $Y = \{a, b, c\}$ is the set of allowed labels. For a path P with vertexes (0, 3, 5, 6), the corresponding edge labels are $\{c, b, a\}$ exactly. In this sense, the path (0, 3, 5, 6) is said to satisfy plc_1.

Fig. 6.3 Labeled directed
graph

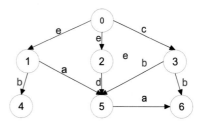

Special Case: NCs

As shown in Fig. 1.1, PLCs extend NCs. For example, nc_1 in Sect. 6.1, can be written a set
of PLCs on single-edge paths as follows:

$$plc_2 : (v, u) \rightarrow \{(\lambda(v), \lambda(u))\}.$$

It states that the only edge $(v, u) \in E$ in the path P has a label from the set $Y = \{(\lambda(v), \lambda(u))\}$
where $(\lambda(v), \lambda(u)) \in N$. As a consequence, PLCs subsume NCs, in other words, PLCs
generalize/extend NCs, denoted by the arrow from NCs to PLCs in Fig. 1.1.

Application

As mentioned in [103], the label-constraint reachability problem has quantities of potential
applications, ranging from social network analysis, viral marketing, to bioinformatics and
RDF graph management. For instance, in a social network graph, relationships between
persons can be presented through labels associated with edges. The label-constraint reachability query can be used to check whether v is reachable from u with respect to the label
constraints. For example, if we want to know whether a person A is related to another person
B, we then can use the query to check if there is a path from A to B with a label-constraint
{parent-of, brother-of, sister-of}.

6.4 XML Functional Dependencies (XFDs)

The eXtensible Markup Language (XML) [27] is known as a standard for data representation
and interchange on the Internet. As a special type of graph data, the XML entry is often in a
tree structure. Functional dependencies have been extended to XML data to improve XML
semantic expressiveness. XML functional dependencies (XFDs) [165] are defined in XML
using the concept of a "tree tuple".

Definition

An *XML Functional Dependency* (XFD) over an XML data tree D is defined as

$$XFD : P_v : X \rightarrow A,$$

where (1) P_v is a downward context path from the root to a considered vertex v, (2) X is a set of entities and (3) A is a single entity. An entity consists of an element name in the XML document and the optional key attribute(s). The XFDs state that for any two instance subtrees identified by P_v, if all the entities in X agree on their values, then they must also agree on the value of A.

Example

Consider an XFD in Fig. 6.4:

$$\text{xfd}_1 : /\text{PSJ}/\text{Project} : (\text{Supplier.SName, Part.PartNo}) \rightarrow \text{Price}.$$

The path P_v indicates that this XFD holds over the subtree rooted at /PSJ/Project. The dependency states that the supplier must supply a part at the same price regardless of project. It is easy to find that the XML in Fig. 6.4 violates the XFD because the supplier "A" sells the same part "B" at different prices "C_1" and "C_2" to two different projects, respectively.

Special Case: FDs

As shown in Fig. 1.1, XFDs subsume FDs in Sect. 1.1. Suppose that Table 1.1 is represented as an XML tree with root *hotel*, as shown in Fig. 6.5. Then the example fd_1 in Sect. 1.1 can be represented by an XFD as follows:

$$\text{xfd}_2 : /root/hotel : \text{address} \rightarrow \text{region}.$$

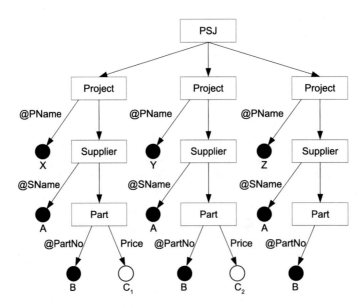

Fig. 6.4 An XML document

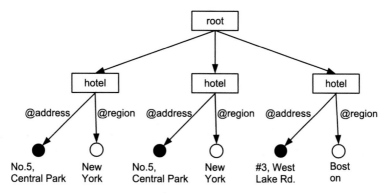

Fig. 6.5 An XML tree from Table 1.1

It means that there should be a same region with the same address. In this sense, XFDs extend/generalize FDs, denoted by the arrow from FDs to XFDs in Fig. 1.1.

Discovery

Yu and Jagadish [182] present an implementation of the XFDs discovery system, *DiscoverXFD*, to discover minimal XFDs efficiently. They introduce two data structures: (1) *attribute partition* is a set of partition groups containing all tuples sharing the same values at X, and (2) *attribute set lattice* represents all FDs. The discovery algorithm traverses the lattice, which is simulated with queue, and discovers the satisfied minimal FDs by constructing and comparing the attribute partitions.

Application

Arenas and Libkin [7] study the normal forms of XML schema and propose a generalization of relational BCNF, referred to as XNF. The definition of XNF and the algorithms for lossless decomposition into XNF are based on XFDs. Similar to FDs in relational databases, XFDs can be used for XML query optimization by rewriting the query according to a set of given constraints. Moreover, XFDs are also useful for data cleaning. Yu and Jagadish [182] use XFDs to discover data redundancies in XML databases. While Flesca et al. [80] suggest that the repair under XML constraints may yield an effective technique for cleaning integrated XML data.

6.5   XML Conditional Functional Dependencies (XCFDs)

Similar to conditional functional dependencies that extend FDs with conditions, XML Conditional Functional Dependencies (XCFDs) [166] are defined. It improves the capability of XFDs by capturing conditional semantics that apply to some fragments rather than the entire XML tree.

Definition

An *XML Conditional Functional Dependency* (XCFD) has a form of

$$\text{XCFD} : P_v : (C, X) \rightarrow A,$$

where P_v is a downward context path from the root to a considered vertex v, X is a set of entity, A is a single entity and C is a condition for the XFD. XCFDs mean that for any two instances of subtrees identified by P_v, if all entities in X with condition C agree on their values, then they must agree on the value of A.

Example

Similar to the example of XFDs in Sect. 6.4, we can consider an XCFD defined in Fig. 6.4,

$$\text{xcfd}_1 : /\text{PSJ}/\text{Project} : (\text{Supplier.SName} = \text{"A"}, \text{Part.PartNo}) \rightarrow \text{Price}$$

Path P_v means that this XCFD holds over the subtree rooted in /PSJ/Project. The supplier named "A" must supply a part at the same price regardless of project. The example in Fig. 6.4 also violates the XCFDs because the supplier "A" sells part "B" at two prices of "C_1" and "C_2".

Special Case: XFDs

Similar to other conditional data dependencies such as CFDs extend FDs, and CDDs extend DDs, XCFDs subsume XFDs as shown in Fig. 1.1. The xfd_1 in Sect. 6.4 can be rewritten by an XCFD as follows:

$$\text{xcfd}_2 : /\text{PSJ}/\text{Project} : (\text{Supplier.SName} = \_, \text{Part.PartNo}) \rightarrow \text{Price}.$$

The condition "_" here means any value of Supplier.SName. In this sense, XCFDs subsume XFDs, in other words, XCFDs extend/generalize XFDs, denoted by the arrow from XFDs to XCFDs in Fig. 1.1.

Discovery

To improve data consistency, Vo et al. [166] propose an approach XDiscover to discover a set of minimal XML Conditional Functional Dependencies from a given XML instance. The algorithm includes four steps. First, the partition identifiers candidate XCFDs are generated. Second, partitions of partition identifiers, associated with each candidate XCFD, are generated. Third, generated partitions are applied to validate for a satisfied XCFD. Fourth, the algorithm will search for minimal XCFDs and prune redundant candidate XCFDs. XDiscover also incorporates a set of pruning rules in discovery process to improve searching performance. The purpose is to improve the search performance by reducing lattice search and the number of XCFD candidates to be checked on the dataset.

Application

Similar to XFDs as well as other integrity constraints, to constrain the data process and minimize the data inconsistency, we can embed XCFDs as an integral part in enterprise's systems. XCFDs can detect and correct non-compliant data.

6.6   Keys for Graph (GKs)

Similar to keys in relational data, a special type of functional dependencies, keys for graphs (GKs) are also introduced to uniquely identify the real-world entities represented by vertexes in a graph [60]. Identifying equivalence among real-world entities could bring applications such as knowledge fusion and identical role detection in social networks. GKs consider the isomorphic subgraph pattern involving the entity. The identical entity detection is based on the isomorphic matching between the subgraph pattern declared by GK and the candidate graph such as knowledge base or social network.

Definition

For each class C of real-world entities represented by vertexes in graph $G = (V, E, L)$, a *key for graph* (GK) considers a subgraph pattern

$$GK : Q(x) = (V_Q, E_Q, L)$$

that uniquely identify each entity in class C, with $V_Q \subseteq V$, $E_Q \subseteq E$ and $x \in V_Q$ as a vertex in subgraph pattern Q of class C. An example could be seen in Fig. 6.6 that states the identification of entity class Album is decided by GK : $Q(x)$ in graphs such as knowledge base.

The semantics of GKs is based on the isomorphic graph matching that finds bijections $h : Q(x) \leftrightarrow G_Q, G_Q \subseteq G$ for pattern $Q(x)$. For any two isomorphic matchings h_1, h_2 of $Q(x)$, when all pairs of vertexes $\langle h_1(v), h_2(v)\rangle$, $v \in V_Q$ except x correspond to the same entity in the real world, it is intuitive that two matched vertexes $\langle h_1(x), h_2(x)\rangle$ are two representations in G of the same entity from real world. Besides, as it is hard for determining

Fig. 6.6 Pattern $Q(x)$ of key for graph

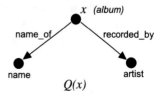

Fig. 6.7 Example G of key for graph

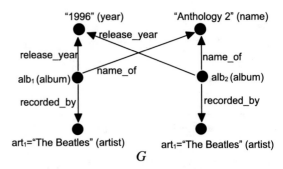

$$G$$

whether two vertexes $\langle h_1(v), h_2(v)\rangle$, $v \in V_Q$ are the same in the real world, key for graph (GK) also considers other types of matchings between $h_1(Q(x))$ and $h_2(Q(x))$ that allows both value and class equivalence on $\langle h_1(v), h_2(v)\rangle$, $v \in V_Q$ to decide whether $\langle h_1(x), h_2(x)\rangle$ correspond to the same entity.

Example

The key for graph $gk_1 : Q(x)$ in Fig. 6.6 considers the identification of the entity class Album in graph G in Fig. 6.7. The constraint $gk_1 : Q(x)$ states that the value matches of vertexes name and artist imply the entity matches of real-world entities. That is, the same artist and the same name are sufficient to identify an album in the real world. Two isomorphic matches of $Q(x)$ could be detected in graph G with bijection h_1, h_2, where $h_1 = \{(x, alb_1), (name, \text{“Anthology 2”}), (artist, art_1)\}$ and $h_1 = \{(x, alb_2), (name, \text{“Anthology 2”}), (artist, art_2)\}$. When the two artists art_1 and art_2 have the same name, e.g., “The Beatles”, the two albums, $h_1(x) = alb_1$ and $h_2(x) = alb_2$, correspond to the same album (an entity) in the real world.

Application

Fan et al. [60] study entity matching problem using keys for graphs, which is to find all pairs of entities in a given graph that can be identified by a set of keys. They develop a MapReduce algorithm and a vertex-centric algorithm to solve entity-matching problem, which is an NP-hard problem. Both of the algorithms are parallel scalable. For emerging applications such as knowledge fusion and knowledge base expansion, deduplication entities, and the fusion of information from different sources from the same entity, GKs are also important. In addition, GKs can be applied into social network reconciliation, which is used to reconcile user accounts across multiple social networks.

6.7   Graph-Patterns Association Rules (GPARs)

Analogous to GKs, GPARs [77] also capture graph patterns to define the constraints on graph data. Compared to GKs, which only support entity identification, GPARs can consider more interesting edge literals, such as "friend" and "visit". By studying association rules between entities, GPARs can be utilized for marketing, social recommendation, linking prediction, etc.

Definition

A *graph-pattern association rule* (GPAR) is defined as

$$\text{GPAR} : Q(x, y) \Rightarrow q(x, y)$$

where $Q(x, y)$ is a graph pattern, x and y are designated nodes, and $q(x, y)$ denotes the label of the edge between x and y. A GPAR states that for any pair of nodes $v_x, v_y \in G$, if there exists a match $h \in Q(G)$ such that $h(x) = v_x$ and $h(y) = v_y$, then the edge $(v_{x,y})$ with label q should hold.

Example

Consider an example of social recommendation. The association rule in Fig. 6.8 can be expressed as a GPAR:

$$\text{gpar}_1 : Q(x, y) \Rightarrow \text{follow}(x, y).$$

where the pattern says that if x and z follow each other, y and z follow each other and y follows x, then x should also follow y. As shown in Fig. 6.9, we can find John and Allen follow each other, and Mike and Allen follow each other, and Mike also follows John. Then we can recommend John to follow Mike.

Fig. 6.8 Pattern $Q(x)$ in GPAR

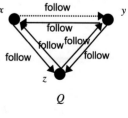

Fig. 6.9 An example G of GPAR

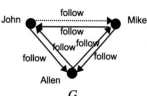

Discovery

Fan et al. [77] study the problem of discovering top-k diversified GPARs, which is NP-hard. Given a graph G, a predicate $q(x, y)$, a support bound σ and positive integers k and d, the problem is to find a set L_k of k nontrivial GPARs such that (a) $F(L_k)$ is maximized; and (b) for each gpar $\in L_k$, supp(gpar, G) $\geq \sigma$ and $r(P_r, x) \leq d$. Here, $F(L_k)$ is the objective function, consisting of the confidence measure of a GPAR and the difference measure between GPARs. And $r(Q, x)$ is the radius of Q at node x. To discover GPARs efficiently, a parallel algorithm with accuracy bound is developed.

Application

By describing relations between entities with $q(x, y)$, GPARs can be used in various scenarios. For instance, Fan et al. [77] propose to use GPARs for link prediction as social recommendation, marketing, knowledge graph completion, etc. Specifically, Fan et al. [77] study the potential customer identification problem using GPARs, which is also an NP-hard problem. With the increase in processors, the parallel algorithm developed by Fan et al. [77] guarantees a polynomial speedup over sequential algorithms.

6.8   Functional Dependencies for Graph (GFDs)

Analogous to keys for graphs (GKs), graph functional dependencies (GFDs) [78] study the functional relationship over the attributes of vertices in graphs. Since the graph data does not come with a fixed schema, a GFD uses an isomorphic subgraph pattern as the scope of involved vertices. In this sense, the pattern is more general than the scope of edges in NCs and paths in NLCs and PLCs. Similar to functional dependencies, GFDs assert the equivalence relationship over the attribute values of every two vertices in the subgraph pattern. GFDs could be applied to manage the graph quality problems such as graph consistency checking and repairing [78, 105].

Definition

A *functional dependency for graphs* (GFD) is a form of

$$\text{GFD} : Q[\bar{x}](X \Rightarrow Y),$$

where (1) $Q[\bar{x}] = (V_Q, E_Q, L_Q)$ is a subgraph pattern with $\bar{x} \subseteq V_Q$ as the list of vertices in $(X \Rightarrow Y)$, (2) X and Y are two sets of equivalent literals composed by the attributes of the vertices in $\bar{x}$. Each of the equivalent literals of $\bar{x}$ has the form of either $x.A = c$ or $x.A = y.B$, where $x, y \in \bar{x}$, A, B denote attributes of vertices x, y, respectively, and c is a constant comparable to the values in attribute A.

The semantics of GFDs is based on the isomorphic subgraph matchings of pattern $Q[\bar{x}]$. Given an isomorphic matching $h : Q \leftrightarrow G_Q, G_Q \subseteq G$ for a GFD : $Q[\bar{x}](X \Rightarrow Y)$, when the set of matched vertices $h(\bar{x})$ satisfies the condition declared by the literals in X, the GFD indicates that the literals in Y are also satisfied by vertices $h(\bar{x})$. The consistency of graph G could be checked by a set of GFDs.

Example

A GFD can be defined with the graph pattern illustrated in Fig. 6.10 which depicts a country entity with two distinct capitals:

$$gfd_1 : Q[x, y, z](x.\text{country} = \text{"USA"} \Rightarrow y.\text{city} = z.\text{city}).$$

It is to ensure that for an entity x, if $x = $ "USA" and has two capital entities y and z, then y and z share the same name. The graph in Fig. 6.11 violates gfd_1, since with $x.\text{country} = $ "USA", it has two different capital cities, i.e., $y.\text{city} = $ "Washington" and $z.\text{city} = $ "New York".

Special Case: XCFDs

From Fig. 1.1 we can find that GFDs extend XCFDs in Sect. 6.5. As shown in Fig. 6.12, we can represent the example $xcfd_1$ in Sect. 6.5 as follows,

$$gfd_2 : Q[x, y, z](x.\text{Supplier.SName} = \text{"A"}, y.\text{Part.PartNo} = z.\text{Part.PartNo}$$
$$\Rightarrow y.\text{Price} = z.\text{Price}).$$

Consider /PSJ/Project as a class. The gfd_2 states that, for an entity x of Supplier whose $x.\text{Supplier.SName}$ is "A", if it has two Parts y and z, and $y.\text{Part.PartNo} = z.\text{Part.PartNo}$, then $y.\text{Price}$ should be equal to $z.\text{Price}$. As a consequence, GFDs subsume FDs, in other words, GFDs extend/generalize FDs, denoted by the arrow from FDs to GFDs.

Fig. 6.10 Graph Q pattern in GFD

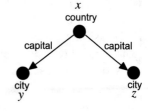

Fig. 6.11 An example G of GFD

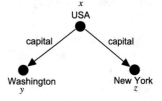

Fig. 6.12 Graph pattern Q in
GFD to represent XCFD

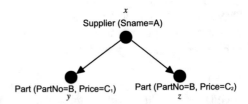

Discovery

Fan [59] studies the discovery of GFDs which is nontrivial. Fan et al. [69] develop parallel scalable algorithms to discover GFDs. There is a study about fixed-parameter tractability of three basic problems related to GFD discovery. It is proved that the implication and satisfiability problems are fix-parameter tractable. They introduce the concept of reduced GFDs and its topological support, and formalize the problem of GFDs discovery. Algorithms are developed to discover GFD and calculate its coverage. GFD discovery is feasible on a large-scale graph. By providing a parallel and scalable GFD discovery algorithm, the running time can be reduced when using more processors.

Application

Fan et al. [78] study the validation problem as an application of GFDs, and prove that it is NP-complete. It is to detect inconsistencies and errors in graphs by using GFDs as data quality rules. They propose an approximate algorithm that is parallel scalable, which makes it feasible to detect errors in large-scale graphs.

The approach proposed by Kang and Wang [105] provides a structure of utilizing a set of GFDs on edge labels to repair the dynamic graph database. On noticing quality problem during the update of knowledge base with RDF triples extracted from corpus streams, Kang and Wang [105] find it difficult to repair graphs with GFDs and discover new GFDs according to the updated graph. To process the update and graph repair dynamically, Kang and Wang [105] consider a subset of GFDs, namely positive and negative graph patterns. The conflict repair is based on the modification of edge label in each additional RDF tuple with the supervision from linkage predication and dynamically updated GFDs.

6.9 Graph Entity Dependencies (GEDs)

While GKs consider the identify semantics of keys in Sect. 6.6, GFDs study the determination of attribute values in Sect. 6.8. To combine both features, graph entity dependences (GEDs) [73] consider the functional relationship among attribute values and constants for consistency checking, and allow the literals with identifier id to identify real-world entities as an extension to KGs. Similar to GFDs, a GED is defined based on a subgraph pattern to introduce the scope of vertices and edges involved in a functional dependency. With the tolerance on the way of graph matching, GED finds matches of its subgraph pattern from the graph through homomorphic matching. GEDs could be applied to inconsistencies discovery in graphs and spam detection Fan and Lu [73].

Definition

The semantics of GEDs could be extended from GFDs in Sect. 6.8 and GKs in Sect. 6.6 by applying homomorphic subgraph matching from pattern Q to a subgraph in graph G.
   A *graph entity dependency* (GED) is defined as

$$\text{GED} : Q[\bar{x}](X \Rightarrow Y)$$

where $Q[\bar{x}]$ is a graph pattern, and X and Y are two sets of *literals* of $\bar{x}$, $X \Rightarrow Y$ is the FD of GED. A *literal* of $\bar{x}$ is one of the following: for $x, y \in \bar{x}$, (1) $x.A = c$ where c is a constant, and A is an attribute; (2) $x.A = y.B$ where A and B are attributes that are not id, which is a special attribute of a vertex, denoting vertex identity; (3) $x.\text{id} = y.\text{id}$.

Example

Figure 6.13 shows a pattern Q for a GED, which is defined as

$$\text{ged}_1 : Q[x, y, x', y'], x.\text{release} = \text{``Apple''}, y.\text{name} = y'.\text{name} \Rightarrow x.\text{id} = x'.\text{id}.$$

In the pattern $Q[x, x', y, y']$ of ged_1, x, x' represent two vertices with class album, and the other two y, y' are with class name. It states that the albums with the same name correspond to the same entity in the real world if one of the albums is released by "Apple", a basic attribute of class album. For the graph in Fig. 6.7, the homomorphic matching $h : Q \rightarrow G_Q, G_Q \subseteq G$ of GED ged_1 could be detected that shows the connections from $Q[x, y, x', y']$ to a subgraph G_Q of G. The mappings of h are represented by $\{(x, \text{alb}_1), (x', \text{alb}_2), (y, \text{``Anthology 2''}), (y', \text{``Anthology 2''})\}$, where the match h finds the same vertex for both y, y' in ged_1. When the attribute value $\text{alb}_1.\text{release} = \text{``Apple''}$, the homomorphic matching h shows satisfaction with respect to the left-hand side of ged_1. Accordingly, we could infer that two matched vertices $h(x) = \text{alb}_1$ and $h(x') = \text{alb}_2$ correspond to the same entity in the real world. The semantics shown by ged_1 indicates that

Fig. 6.13 Pattern Q of a GED

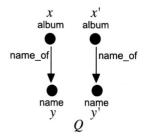

the constraints of GED could handle more tasks for graph data quality. As follows, we use examples to briefly discuss the special cases of GEDs, i.e., GFDs and GKs.

Special Case: GFDs

GEDs subsume GFDs in Sect. 6.8 as special cases. GFDs are syntactically defined as GEDs without id literal. For example, gfd_1 in Sect. 6.8 can be regarded as a GED since there is not id literal,

$$ged_2 : Q[x, y, z], x.\text{country} = \text{"USA"} \Rightarrow y.\text{city} = z.\text{city}).$$

It states that with a country as "USA", we must have $y.\text{city} = z.\text{city}$, since a country can only have one capital. In this sense, GEDs subsume GFDs, i.e., GEDs extend/generalize GFDs as denoted in Fig. 1.1 with an arrow from GFDs to GEDs.

Special Case: GKs

GEDs subsume GKs in Sect. 6.6 as special cases. A key for graph is defined as a GED of the form $Q[x, y, \dots](X \rightarrow x.\text{id} = y.\text{id})$. For example, we can present the GK in Sect. 6.6 as a GED,

$$ged_3 : Q[x, y](\emptyset \Rightarrow x.\text{id} = y.\text{id}).$$

It means that all vertexes matching the pattern Q must have the same id, i.e., denoting the same entity in the real world. As a consequence, GEDs extend/generalize GKs, denoted by the arrow from GFDs to GKs in Fig. 1.1.

Axiomatization

For the class of graph data, GEDs are finitely axiomatizable and the set of inference rules $\mathcal{A}_{GED}$ is sound, complete and independent (non-redundant and minimal) [73]. $\mathcal{A}_{GED}$ contains six inference rules. Let Eq_X be the equivalence relation of a set X of literals in G_Q and $(G_Q)_{\text{Eq}}$ be the coercion of Eq on G_Q.

GED1 : $\Sigma \vdash Q[\bar{x}](X \rightarrow X \wedge X_{\text{id}})$, where X_{id} is $\bigwedge_{i \in [1,n]}(x_i.\text{id} = x_i.\text{id})$, and $\bar{x}$ consists of x_i for all $i \in [1, n]$.

GED2 : If $\Sigma \vdash Q[\bar{x}](X \rightarrow Y)$, and literal $(u.\text{id} = v.\text{id}) \in Y$, then $\Sigma \vdash Q[\bar{x}](X \rightarrow u.A = v.A)$ for all attributes $u.A$ that appear in Y.

GED3 : If $\Sigma \vdash Q[\bar{x}](X \rightarrow Y)$ and $(u = v) \in Y$, then $\Sigma \vdash Q[\bar{x}](X \rightarrow v = u)$.

GED4 : If $\Sigma \vdash Q[\bar{x}](X \rightarrow Y)$, $(u_1 = v) \in Y$ and $(v = u_2) \in Y$, then $\Sigma \vdash Q[\bar{x}]$ $(X \rightarrow u_1 = u_2)$.

GED5 : If $\Sigma \vdash Q[\bar{x}](X \rightarrow Y)$ and $\text{Eq}_X \cup \text{Eq}_Y$ is inconsistent, then $\Sigma \vdash Q[\bar{x}]$ $(X \rightarrow Y_1)$ for any set Y_1 of literals of $\bar{x}$.

GED6 : If $\Sigma \vdash Q[\bar{x}](X \rightarrow Y)$, $\text{Eq}_X \cup \text{Eq}_Y$ is consistent, $\Sigma \vdash Q_1[\bar{x}_1](X_1 \rightarrow Y_1)$, and there exists a match h of Q_1 in $(G_Q)_{\text{Eq}_X \cup \text{Eq}_Y}$ such that $h(\bar{x}_1) \models X_1$, then $\Sigma \vdash Q[\bar{x}]$ $(X \rightarrow Y \wedge h(Y_1))$.

Following Armstrong's axioms [8], GED$_1$ extends Reflexivity to consider id literals. The augmentation rule and transitivity rule can also be derived from $\mathcal{A}_{GED}$.

Application

As indicated in [73], GEDs have great expressive power compared to other data dependencies in graphs. They are defined according to graph patterns as topological constraints and are used to describe the association among entities in a graph. This can make native graph techniques applied to dependencies analysis, for example, the locality of graph homomorphism, which is not considered by traditional relational data dependencies techniques. GEDs are useful in consistency checking, spam detection, entity resolution and knowledge base expansion for graph data. Moreover, they can help in optimizing queries that are costly on large graphs in practice.

6.10   Graph Association Rules (GARs)

While GFDs and GEDs study the functional relationships over the attributes of vertices in schemaless graphs, graph association rules (GARs) [71] further take ML predicates into consideration and can catch both missing links and semantic errors. Analogous to graph functional dependencies (GFDs), a GAR is also defined on a subgraph pattern to introduce the scope of involved vertices. By matching the pattern to the subgraphs in the given graph, GARs can be applied to deduce associations for link prediction, customer identification, etc. [71].

Definition

A *graph association rule* (GAR) φ is defined as

$$\text{GAR} : Q[\bar{x}](X \rightarrow Y)$$

where $Q[\bar{x}]$ is a graph pattern, X and Y are two sets of *literals* of $\bar{x}$, $X \rightarrow Y$ is the dependency of φ. A *literal* of $\bar{x}$ is one of the following: for $x, y \in \bar{x}$ and attributes A, B, (1) attribute literal $x.A$; (2) edge literal $\iota(x, y)$, where ι denotes the edge label; (3) ML literal $\mathcal{M}(x, y, \iota)$, an ML classifier that returns true if and only if it predicates the edge (x, ι, y) exists; (4) constant literal $x.A = c$, where c is a constant; (5) variable literal $x.A = y.B$.

Analogous to GFDs, the semantics of GARs are also based on the isomorphic subgraph matching of pattern $Q[\bar{x}]$. Given an isomorphic matching h for a GAR, when the set of matched vertices $h(\bar{x})$ satisfies the condition declared by the literals in X, the GAR indicates that the literals in Y also hold. Compared to GFDs, GARs support more literals.

Example

Figure 6.14 shows a pattern Q for a GAR, which is defined as

$$\text{gar}_1 : Q[x, x'](X \rightarrow Y),$$

Fig. 6.14 Pattern Q of a GAR

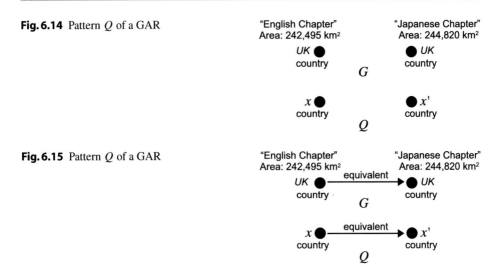

Fig. 6.15 Pattern Q of a GAR

where X is x.name $= x'$.name $\wedge\, M(x, x',$ equivalent$)$, and Y is equivalent(x, x'). It means that if two countries have the same name and the ML classifier predicates that the two countries are equal, then a link (equivalent) between the two countries should be added. In Fig. 6.14, the names of the two countries are "UK", and the ML model predicts that the two countries should be equivalent. Thus, gar_1 can catch the missing link between two nodes.

Special Case: GFDs and GEDs

GARs extend GFDs and GEDs by considering more literals other than variable and constant literals, assuming that id is also a kind of attribute. Consider another example in Fig. 6.15, a gfd : $Q[x, x'](\emptyset \Rightarrow Y)$ can be represented as a GAR,

$$\mathsf{gar}_2 : Q[x, x'](\emptyset \rightarrow Y),$$

where Y in gfd and gar_2 is x.area $= x'$.area. It states that the same countries should have same area and can catch inconsistencies in the example. The inconsistency in G is thus captured by gfd and gar_2, since the areas of UK in the English and Japanese chapters are different. In this sense, GARs subsume the semantics of GFDs/GEDs.

Special Case: GPARs

Graph pattern association rules (GPARs) [77] can be expressed as GARs $Q[\bar{x}](\emptyset \rightarrow \iota(x, y))$, where no precondition is specified in X and Y is an edge literal. For instance, gpar_1 : $Q(x, y) \Rightarrow \mathsf{follow}(x, y)$ can be expressed as a GAR,

$$\mathsf{gar}_3 : Q[x, y, z](\emptyset \rightarrow \mathsf{follow}(x, y)).$$

In contrast to GARs, GPARs do not embed ML classifiers.

Discovery

Fan et al. [71] propose to discover GARs by extending GFDs discovery algorithm [68]. (1) All missing links predicated by the ML classifier will be added as edges. (2) Following [68], starting from the frequent nodes, the algorithm extends patterns and finds attribute dependencies by interleaving vertical and horizontal spawning. To obtain additional literals in GARs, some edges in the discovered pattern are added as edge literals in GARs. The attributes in the matches are included as edge literals. (3) For ML literals, certain edge literals $\iota(x, y)$ are replaced with $\mathcal{M}(x, y, \iota)$.

To reduce the discovery cost of GARs from big graphs, Fan et al. [61] present an application-driven strategy. The strategy removes the irrelevant data identified by the ML model, reducing the original graph G to a smaller graph $G_{\mathcal{A}}$. To further accelerate the discovery, Fan et al. [61] propose to discover GARs from a sampled set H, consisting of representative data cells along with the corresponding subgraphs in $G_{\mathcal{A}}$. Fan et al. [61] provide theoretical bounds of recall and support for the rules discovered from sampled H. Furthermore, a parallel algorithm is developed and guarantees the parallel scalability [61].

Application

As indicated in [71], GARs extend GFDs to capture missing links and inconsistencies. Thus, GARs may not only be used to all applications of GFDs, but also to the applications related to link prediction, such as knowledge graph completion, recommendation, and so on [61].

6.11 Graph Differential Dependencies (GDDs)

Since GEDs only support exact match over the attributes of vertices in graphs, GEDs cannot capture the semantics of approximate matches. To meet the need of approximate matches in data quality applications, Kwashie et al. [114] propose graph differential dependencies (GDDs) that introduce similarity relationships to graph dependencies. Benefit from the similarity-based matches, GDDs achieve good performance in the entity resolution task [114].

Definition

A *graph differential dependency* (GDD) φ is defined as

$$\text{GDD} : Q[\overline{x}](\Phi_X \rightarrow \Phi_Y)$$

where $Q[\overline{x}]$ is a graph pattern, $\Phi_X \rightarrow \Phi_Y$ is the dependency, Φ_X and Φ_Y are two sets of distance constraints on the pattern variables $\overline{x}$. A distance constraint in Φ_X and Φ_Y is one of the following:

$$\delta_A(x.A, c) \le t_A; \quad \delta_{A_1 A_2}\left(x.A_1, x'.A_2\right) \le t_{A_1 A_2};$$
$$\delta_{\equiv}(x.\text{eid}, c_e) = 0; \quad \delta_{\equiv}\left(x.\text{eid}, x'.\text{eid}\right) = 0;$$
$$\delta_{\equiv}(x.\text{rela}, c_r) = 0; \quad \delta_{\equiv}\left(x.\text{rela}, x'.\text{rela}\right) = 0;$$

where $x, x' \in \overline{x}$, A_1, A_2 are attributes in A, $c \in dom(A)$ is value in the domain of attribute A, $\delta_{A_1 A_2}(\cdot, \cdot)$ is a distance function ($\delta_{A_1 A_2}$ can be written as δ_A if $A_1 = A_2$). Here, $\delta_{\equiv}(x.\text{eid}, c_e) = 0$ denotes that the eid of x is c_e. Moreover, $\delta_{\equiv}(x.\text{rela}, c_r) = 0$ means that the node x has a relation named rela that ended with the node c_r. Furthermore, $\delta_{\equiv}\left(x.\text{eid}, x'.\text{eid}\right) = 0$, if x and x' have the same eid. Finally, $\delta_{\equiv}\left(x.\text{rela}, x'.\text{rela}\right) = 0$, if both x and x' have the relation named rela and ended with the same node.

Example

Figure 6.16 shows a pattern Q for a GDD, which is defined as

$$\text{gdd}_1 : Q[x, x'](\Phi_X \to \Phi_Y),$$

where Φ_X is $\delta_{\text{name}}(x, x') \le 1 \wedge \delta_{\text{dept}}(x, x') \le 0$, and Φ_Y is $\delta_{\equiv}(x.\text{eid}, x'.\text{eid}) = 0$. gdd_1 says that any pair of person entities have similar names and from the same department should refer to the same real-world person.

Special Case: GEDs

Compared to GEDs, GDDs consider more literals about relation (edges) such as $\delta_{\equiv}(x.\text{rela}, c_r) = 0$. In addition, GDDs relax the equality relationship (i.e., $t_A = 0$) and capture the approximate matches in the dependency. Thus, when no edges-related literals are considered and all distance threshold are 0, it is exactly a GED. For instance, consider the GED below,

$$\text{ged} : Q[x, x'](x.\text{name} = x'.\text{name} \wedge x.\text{dept} = x'.\text{dept} \Rightarrow x.\text{eid} = x'.\text{eid}).$$

Fig. 6.16 Pattern Q of a GDD

The ged can be represented as a GDD,

$$gdd_2 : Q[x, x'](\delta_{name}(x, x') \leq 0 \wedge \delta_{dept}(x, x') \leq 0 \rightarrow \delta_{\equiv}(x.eid, x'.eid) = 0).$$

It means that two people with the same name and from the same department should be identified as the same person. In this sense, GDDs generalize GEDs, denoted by an arrow in Fig. 6.19.

Discovery

Kwashie et al. [114] investigate the discovery of Linking GDDs (GDD$_L$) whose Φ_Y is $\{\delta_{\equiv}(x.eid, x'.eid) = 0\}$. For every entity type et, to accelerate the discovery process, the algorithm first retrieves the set of all matches of the pattern $Q[x, x']$ in G and the subgraph G_{et}, whose nodes are et-typed or connect to et-typed nodes, of G. Then, the discovery algorithm constructs single itemsets and each itemset contains one (item-name, threshold) pair. With the given threshold levels, the satisfaction sets of itemsets $sat(\mathcal{L})$ can be computed. If $sat(\mathcal{L}) \in sat(\Phi_{eid})$, the itemset $\mathcal{L}$ is called completed and can be taken as Φ_X in the corresponding GDD$_L$. If not, the algorithm derives itemsets for higher levels by pairing the remaining itemsets $\mathcal{L}' = \mathcal{L}_1 \bigcup \mathcal{L}_2$ and the satisfaction set is $sat(\mathcal{L}') = sat(\mathcal{L}_1) \bigcap sat(\mathcal{L}_2)$. The discovery finishes until no itemset is possible.

Application

As indicated in [114], the linked GDDs (GDD$_L$) can be applied for entity resolution. Kwashie et al. [114] develop an ER algorithm based on GDD$_L$. The algorithm allows a low similarity score for high recall and utilizes GDDs to avoid too many false positives for high precision.

6.12 Graph Denial Constraints (GDCs)

In order to further improve the expressive power of GEDs, Fan and Lu [73] propose graph denial constraints, denoted by GDCs, following the same line of DCs that can represent all FDs/CFDs. They replace the equality relationship in the definition of GEDs with built-in predicates $=, \neq, <, >, \leq$ and $\geq$.

Definition

A *graph denial constraint* (GDC) is defined as

$$GDC : Q[x](X \Rightarrow Y),$$

where Q is a graph pattern, X and Y are sets of literals in one of the following forms: (a) $x.A \oplus c$, (b) $x.A \oplus y.B$, for constant $c \in U$, and non-id attributes A, B, and (c) $x.id = y.id$; here $\oplus$ is one of built-in predicates $=, \neq, <, >, \leq, \geq$.

Example

Consider the graph pattern in Fig. 6.6 and graph data in Fig. 6.7. A GDC is declared as follows,

$$gdc_1 : Q[x, y](x.\text{name} = y.\text{name}, x.\text{artist} = y.\text{artist},$$
$$x.\text{release\_year} \neq y.\text{release\_year} \Rightarrow \text{false}).$$

It states that albums with the same name and the same artist should not have different release year. As shown in Fig. 6.7, if the name is "alb_1", the artist is "art_1", then there should be only one release_year of "1996".

Special Case: GEDs

As presented in Fig. 1.1, GEDs are special case of GDCs when $\oplus$ is equality '=' only. The ged_1 in Sect. 6.9 can be regarded as a GDC,

$$gdc_2 : Q[x, y, x', y'](x.\text{release} = \text{``Apple''}, y.\text{name} = y'.\text{name} \Rightarrow x.\text{id} \neq x'.\text{id} \Rightarrow \text{false}).$$

In this sense, GDCs subsume GEDs, in other words, GDCs extend/generalize GEDs as denoted in Fig. 1.1 with an arrow from GEDs to GDCs.

6.13 Temporal Dependencies for Graph (TGFDs)

Graphs are increasingly used to model the information about entities and attributes, as well as their relationships. In practice, many of graphs are dynamic and evolving. The entities are constantly changing in such graphs. For example, relationships between customers, their product purchases, and the relationships between products, are all changing over time. Accurate and complete information are essential for the downstream decision-making and fact checking tasks. Following the same line of TFDs extending FDs in the time dimension, Alipourlangouri [5] proposes Temporal Dependencies for Graphs (TGFDs). The dependency specifies the semantics of temporal graph data to detect inconsistencies.

Definition

A *temporal dependency for graph* (TGFD) is a triple

$$\text{TGFD} : Q[\bar{x}], \Delta, X \rightarrow Y,$$

where (1) $Q[\bar{x}]$ is a graph pattern that helps us to capture the entities in a temporal graph, and $\bar{x}$ consists of all the vertexes in Q; (2) Δ is a time interval, defined as a pair (p, q), $p \leq q$; (3) $X \rightarrow Y$ is the data dependency, with two (possibly empty) sets of literals of $\bar{x}$.

The literals in $\bar{x}$ are in the form $x_{ti}.A = c$ or $x_{ti}.A = y_{tj}.B$, where $x_{ti}, y_{tj} \in \bar{x}$, A and B are attributes, and c is a constant. Similar to GFDs, $x_{ti}.A = c$ is referred as a constant literal, and $x_{ti}.A = y_{tj}.B$ as a variable literal [78]. However, in contrast to GFDs, it associates timestamps t_i, t_j to the variables in $\bar{x}$ to explicitly specify that the dependencies may span across time, i.e., $t_j - t_i < \Delta$. Therefore, $Q[\bar{x}]$ and Δ determine the scope of TGFD, in which the data dependency $X \to Y$ needs to be checked.

Example

As an example, to capture inconsistency in the evolution of a hotel, we can consider a TGFD below

$$\mathsf{tgfd}_1 : Q[p, x, y, z], (5 \text{ years}), (x_{t_i} = x_{t_j} \wedge y_{t_i} = y_{t_j}) \to (z_{t_i} = z_{t_j}),$$

where the graph pattern Q is depicted in Fig. 6.17. The black points represent entities, and white points represent attributes. The tgfd_1 enforces that for every two instances of hotel with time difference in the range of 0 to 5 years, if both the name and address of them are same, then they should also have the same star. Assume that the hotel are rated every five years and the hotel "New Center" is rated in 2018. By enforcing tgfd_1 over the graph in Fig. 6.18, we find the pair p_2 in 2020 and p_3 in 2022 that meets the criteria for Δ but violate tgfd_1, i.e., having the same name and address but a different star.

Special Case: GFDs

From Fig. 1.1 we can find that TGFDs extend GFDs. The GFD in Sect. 6.8 can be represented as a TGFD with time interval as follows,

$$\mathsf{tgfd}_2 : Q[x, y, z], (-\infty, \infty), \emptyset \to y.\mathsf{name} = z.\mathsf{name}.$$

It states that, without any restriction on time, for all country entities x, if x has two capital entities y and z, then y and z share the same name. In this sense, TGFDs extend/generalize GFDs, denoted by the arrow from GFDs to TGFDs in Fig. 1.1.

Application

As mentioned in [5], temporal dependencies for graphs can be used to ensure the accuracy of data. TGFDs are defined as temporal graph data dependencies, which can specify the semantics of both temporal and graph data to detect inconsistency. As a consequence, TGFDs can maintain the quality of data on temporal graphs.

Fig. 6.17 A graph pattern Q to capture the entity p of a type hotel

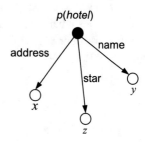

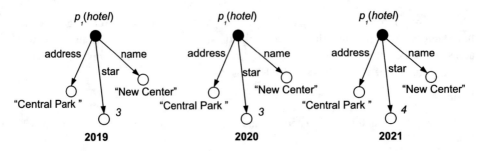

Fig. 6.18 Evolution of a hotel p_1 from 2018 to 2022

6.14   Summary and Discussion

Graph data widely exist in real-world applications such as knowledge bases, workflows or social networks [186]. Like other types of big data, veracity issues are prevalent in graph data. For example, gene ontology annotation could be erroneous in a protein interaction network [153]. The event names are misplaced in a workflow network [169]. To clean such errors, neighborhood constraints between vertexes are considered, e.g., extracted from workflow specification [164]. Unlike relational data, graph data usually do not have a specific schema. Thereby, most of the existing data dependencies cannot be applied in graph data directly. Graph data have structural characteristics that can be used for defining dependencies [73], while they also have various forms, such as XML, RDF or networks. Each form of graph may have its own features. The variety of graph data is another challenge when solving problems in data [60, 78].

The integrity constraints extended for graphs are also useful in applications. For example, in RDF data [187], data dependencies on link information could be interestingly specified and utilized in keyword search optimization. Referring to Table 6.1 and Fig. 6.19, we summarize the relationships among the integrity constraints over graph data in different (sub)categories as follows.

Path

To specify constraints on graphs, a natural idea is to consider the neighborhood of vertexes, in particular, whether the labels can appear in adjacent vertexes. NCs [153] specify the label pairs allowed to appear on adjacent vertexes in the graph, and is used to detect and repair wrong vertex labels and graph structures. Such a constraint on the edges of two vertexes can be further extended to a path. While NLCs [28] consider the allowed labels on the vertexes in a path, PLCs [103] study the labels that should appear on the edges in a path.

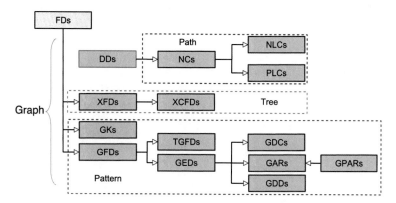

Fig. 6.19 A family tree of extensions between graph data dependencies

Tree

For the data in a tree structure, such as XML, the constraints can be declared over the (sub)trees, in addition to paths. Similar to FDs, XFDs [165] are defined in XML using the concept of a "tree tuple", i.e., some attribute values may determine others in the tree. Likewise, following CFDs that extend FDs with conditions, XCFDs [166] improve the capability of XFDs. It captures conditional semantics that apply to some fragments rather than the entire XML tree.

Pattern

Finally, the constraints are extended from simple paths and trees to more complicated patterns. Again, (temporal) functional dependencies or even denial constraints can be declared over the subgraphs identified by the patterns. GKs [60] are a class of keys for graphs to uniquely identify entities represented by vertexes in a graph. Keys are fundamental to graph for their great effects on data models, conceptual design, and prevention of update anomalies. GFDs [78] capture both attribute-value dependencies and topological structures of entities, to detect further inconsistencies in graphs. GEDs [73] are also a combination of graph pattern and an attribute dependency. In a uniform format, GEDs express GFDs with constant literals to catch inconsistencies, and GKs to identify entities in a graph. Moreover, analogous to DCs that can represent all FDs/CFDs, GDCs [73] further extend GEDs with more built-in predicates, $=$, $\neq$, $<$, $>$, $\leq$ and $\geq$. Finally, following TFDs, TGFDs [5] extend GFDs in the time dimension, i.e., GFDs hold only for a period of time.

Conclusions and Directions

Data dependencies are important in database theory and often used in database design. Recently, data dependencies have been revisited and extended for data tasks, such as error detection and data deduplication, over rich data types. In this study, we give an entire landscape of typical data dependencies, in order to identify their relationships and distinct application scenarios.

Categorical Data

Data dependencies for categorical data are often defined on the equality relationships. Owing to the variety of issues in big data, data dependencies may not exactly hold. We introduce the extensions of data dependencies that are still based on equality relationships. We categorize the extensions into four subcategories: Statistical, Conditional, Multivalued and Inclusion. It is worth noting that most of these extensions are still based on the equality relationship of data values, i.e., not effective in addressing the various information formats on heterogeneous data.

Heterogeneous Data

Data obtained from merging heterogeneous sources often have various representation conventions. Carefully declared data dependencies over the heterogeneous data would be useful in query optimization and consistent query answering. We further categorize the data dependencies over heterogeneous data in three aspects: Metric, Additional and Identification. In this way, we summarize and compare the extension relationships between dependencies.

Order Data

Data sets with ordering relationships are prevalent, e.g., timestamps, sequence numbers, sales, temperature, stock prices, and so on. We introduce several typical studies on such ordered data and summarize the extension relationships between them. We categorize these

© The Author(s), under exclusive license to Springer Nature Switzerland AG 2023   131
S. Song and L. Chen, *Integrity Constraints on Rich Data Types*,
Synthesis Lectures on Data Management,
https://doi.org/10.1007/978-3-031-27177-9_7

dependencies into two classes: Order and Distance. Compared to the dependencies from class Order, the dependencies from class Distance further specify the amount of value change.

Temporal Data

Temporal data are prevalent in industry, often large and dirty. The temporal data quality issues are unique in challenges due to the presence of autocorrelations, trends, seasonality, gaps and expected to be explained. To solve the problems of data management and data quality, various types of data dependencies on temporal data are proposed. We introduce several typical works over temporal data and categorize them into three subcategories: Relational, Timeseries and Event. Again, the extension relationships are summarized and compared.

Graph Data

Graph data have been widely observed in real-world applications, e.g., knowledge bases and social networks can be modeled as graphs. Unlike relational data, graph data do not have a specific schema. In this sense, novel types of integrity constraints need to be developed for graph data. We present the major data dependencies extended for graph data. The corresponding special cases, where the data dependencies are extended from, are presented as well. Furthermore, we categorize these constraints on graph data into three subcategories: Path, Tree and Pattern.

Directions

Despite extensions of data dependencies in recent decades, there are still many problems that have not been fully studied. First, most data dependencies focus on equality relationships and simple similarity relationships, without capturing semantic similarity. Machine learning community has made tremendous progress in semantic understanding, such as natural language processing and knowledge graph inference. It is interesting to embed machine learning rules in data dependencies to enhance their expressive ability. Second, data dependencies on many other data types may still be absent, such as uncertain data. Third, in terms of extension relationships, there are still several parts of data dependencies disjoint with others in the family tree, with unknown connections. Fourth, many data dependencies still lack axiomatization. A complete and sound inference rules (axioms) are important for rule discovery and symbolic proofs. Finally, many data dependencies still need to be defined by experts, and there is no efficient automatic discovery algorithm. To provide users with out-of-the-box data tools to take advantage of mass data, the need for efficient discovery algorithms may be urgent. We envision several emergent research directions, hoping to promote the research and application of data dependencies.

Machine Learning Rules

Bertossi and Geerts [19] propose some ideas about the role of explanation in data quality in the context of data-based machine learning models. It discusses how explainable AI brings a new perspective to data quality in several aspects, especially, causality and data dependencies in databases. In addition to using integrity constraints as data quality rules and matching rules [62, 70], machine learning rules are also studied in [71, 74, 184]. Among the machine learning approaches, Boolean functions take the place of traditional logic predicates to construct the rules. As logic predicates $A\phi B$ can be treated as special linear Boolean classifier, the differences in semantics between machine learning rules and traditional logic rules are mainly caused by the complexity of machine learning functions. Fan et al. [71, 74] apply machine learning predicates to achieve tasks such as linkage prediction in graph constraints. Zhang et al. [184] notice that using machine learning functions for each tuple could generate accurate and effective models for attribute value prediction and repairs. Kang et al. [104] extend the idea to the regression models that conditionally apply to part of the data. Although more complicated semantics could be expressed by extending logic predicates to machine learning models, it is always hard to consider the implementation problems when rules are combined by both kinds of predicates. More studies are highly demanded to analyze more potential machine learning predicates and how to achieve sound logic systems when machine learning models are introduced.

Uncertain Data

Uncertain data are prevalent [141], e.g., owing to noises that make it deviate from the correct or original value in sensor networks [35]. The data are expected to be cleaned by reducing the uncertainty [34]. Motivated by the probability measures of functional dependencies in Sect. 2, a natural idea is to study the data dependencies over uncertain data. An uncertain relation allows to give multiple possible values for tuples, and represents a set of possible worlds, each of which is an ordinary relation. Sarma et al. [141] study the functional dependencies for uncertain relations. The proposed horizontal FDs and vertical FDs are consistent with the conventional FDs when an uncertain relation does not contain any uncertainty. Xiang et al. and Lian et al. [120] study the problem of consistent query answering over uncertain data. A novel notion of repaired possible worlds (involving both repair and possible worlds) is proposed. In this sense, uncertainty could be introduced in all aspects, naturally embedded in the data, probed in the discovered dependencies, or generated in possible repairs. It is highly demanded to holistically study these aspects of uncertainty.

Extensions of Dependencies

As illustrated in Fig. 1.1, several parts are disjoint in the family tree, especially in the type of temporal data. Connecting disjointed parts of the family tree may inspire researchers to find interesting extensions of new data dependencies. For example, TFDs [1, 102, 178] enhance temporal constraints. In this way, TFDs can restrict the event-type attributes with the time dimension and can be used for identifying temporal outliers [1]. Alipourlangouri [5] proposes TGFDs that extend GFDs by considering time intervals. TGFDs can specify the semantics of temporal graph data where entities as well as their relationships are changing over time. From another perspective, we may consider patterns in data dependencies over temporal data. Analogous to graph data, temporal data may not have a complex schema. However, there are also some patterns in temporal data, such as periodicity. Following GFDs [78], an interesting idea is whether patterns could be introduced in place of schema in dependencies on temporal data.

Axiomatization

A complete and sound set of inference rules are useful for rule discovery and symbolic proofs. Unfortunately, many data dependencies, such as SDs, CSDs, GARs [71, 87], do not have the corresponding axioms yet. An interesting direction is to study axiomatization for these dependencies. To meet the need for data analysis, sometimes data scientists may consider multiple kinds of data dependencies at the same time. For instance, ODs and DDs may be used together to verify the correlation between attributes in the meantime. There may not exist any conflict within each type of dependencies, since they are discovered or specified by domain experts independently. However, when multiple types of dependencies are considered at the same time, inconsistencies may arise. Therefore, before applying dependencies, the first step is to ensure the consistencies of them, especially those given by users. To reduce the workload of data scientists, automated validation is necessary, which needs to be supported by axiomatization. Thereby, another direction is to study the inference rules for integrating multiple types of data dependencies.

Discovery

For big data analysis, it is too costly if data dependencies have to be defined by domain experts for each dataset. Efficient discovery algorithms are the key to the wide application of data dependencies. However, up to now, there is still a lack of discovery algorithms for many data dependencies, such as SDs [87] and TGFDs [5]. How to design efficient discovery algorithms for these dependencies is an interesting problem. Besides the conventional pruning techniques, sampling, parallelism [61] and deep learning methods [131] are recently introduced to speed up the discovery of rules. How to transfer these techniques to the discovery of other data dependencies is also worth investigating.

Index of Data Dependencies

Abbreviation	Data dependency	Section
ACs	Acceleration Constraints	5.5
AFDs	Approximate Functional Dependencies	2.5
AINDs	Approximate Inclusion Dependencies	2.13
AMVDs	Approximate MVDs	2.11
BODs	Band Order Dependencies	4.3
CDDs	Conditional Differential Dependencies	3.4
CDs	Comparable Dependencies	3.5
CFDs	Conditional Functional Dependencies	2.7
CINDs	Conditional Inclusion Dependencies	2.14
CMDs	Conditional Matching Dependencies	3.10
CSDs	Conditional Sequential Dependencies	4.6
DCs	Denial Constraints	4.3
DDs	Differential Dependencies	3.3
eCFDs	extended CFDs	2.8
EDGs	Equality Generating Dependencies	2.2
FDs	Functional Dependencies	2.1
FFDs	Fuzzy Functional Dependencies	3.7
FHDs	Full Hierarchical Dependencies	2.10
GARs	Graph Association Rules	6.10
GDDs	Graph Differential Dependencies	6.11
GDCs	Graph Denial Constraints	6.12
GEDs	Graph Entity Dependencies	6.9
GFDs	FDs for Graphs	6.8
GKs	Keys for Graphs	6.6

S. Song and L. Chen, *Integrity Constraints on Rich Data Types*,
Synthesis Lectures on Data Management,
https://doi.org/10.1007/978-3-031-27177-9

Abbreviation	Data dependency	Section
GPARs	Graph-Patterns Association Rules	6.7
INDs	Inclusion Dependencies	2.12
MDs	Matching Dependencies	3.9
MFDs	Metric Functional Dependencies	3.1
MSCs	Multi-Speed Constraints	5.4
MVDs	Multivalued Dependencies	2.9
NCs	Neighborhood Constraints	6.1
NEDs	Neighborhood Dependencies	3.2
NLCs	Node Labels Constraints	6.2
NUDs	Numerical Dependencies	2.6
ODs	Order Dependencies	4.2
OFDs	Ordered Functional Dependencies	4.1
ONFDs	Ontology Functional Dependencies	3.8
PACs	Probabilistic Approximate Constraints	3.6
PFDs	Probabilistic Functional Dependencies	2.4
PLCs	Path Label Constraints	6.3
PNs	Petri Nets	5.7
SCs	Speed Constraints	5.3
SDs	Sequential Dependencies	4.5
SFDs	Soft Functional Dependencies	2.3
TCs	Temporal Constraints	5.6
TDs	Trend Dependencies	5.2
TFDs	Temporal Functional Dependencies	5.1
TGFDs	Temporal Dependencies for Graphs	6.13
XCFD	XML Conditional Functional Dependencies	6.5
XFD	XML Functional Dependencies	6.4

References

1. Abedjan, Z., Akcora, C. G., Ouzzani, M., Papotti, P., and Stonebraker, M. (2015). Temporal rules discovery for web data cleaning. *PVLDB*, 9(4):336–347.
2. Abedjan, Z., Golab, L., Naumann, F., and Papenbrock, T. (2018). *Data Profiling*. Synthesis Lectures on Data Management. Morgan & Claypool Publishers.
3. Abiteboul, S., Hull, R., and Vianu, V. (1995). *Foundations of Databases*. Addison-Wesley.
4. Alexe, B., Roth, M., and Tan, W. (2014). Preference-aware integration of temporal data. *Proc. VLDB Endow.*, 8(4):365–376.
5. Alipourlangouri, M. (2021). Temporal dependencies for graphs. In *SIGMOD '21: International Conference on Management of Data, Virtual Event, China, June 20-25, 2021*, pages 2881–2883.
6. Arenas, M., Bertossi, L. E., and Chomicki, J. (1999). Consistent query answers in inconsistent databases. In *Proceedings of the Eighteenth ACM SIGACT-SIGMOD-SIGART Symposium on Principles of Database Systems, May 31 - June 2, 1999, Philadelphia, Pennsylvania, USA*, pages 68–79.
7. Arenas, M. and Libkin, L. (2004). A normal form for XML documents. *ACM Trans. Database Syst.*, 29:195–232.
8. Armstrong, W. W. (1974). Dependency structures of data base relationships. In *IFIP Congress*, pages 580–583. North-Holland.
9. Baskaran, S., Keller, A., Chiang, F., Golab, L., and Szlichta, J. (2017). Efficient discovery of ontology functional dependencies. In *Proceedings of the 2017 ACM on Conference on Information and Knowledge Management, CIKM 2017, Singapore, November 06 - 10, 2017*, pages 1847–1856.
10. Bassée, R. and Wijsen, J. (2001). Neighborhood dependencies for prediction. In *Knowledge Discovery and Data Mining - PAKDD 2001, 5th Pacific-Asia Conference, Hong Kong, China, April 16-18, 2001, Proceedings*, pages 562–567.
11. Bauckmann, J., Abedjan, Z., Leser, U., Müller, H., and Naumann, F. (2012). Discovering conditional inclusion dependencies. In *21st ACM International Conference on Information and Knowledge Management, CIKM'12, Maui, HI, USA, October 29 - November 02, 2012*, pages 2094–2098.

© The Editor(s) (if applicable) and The Author(s), under exclusive license to Springer Nature Switzerland AG 2023
S. Song and L. Chen, *Integrity Constraints on Rich Data Types*,
Synthesis Lectures on Data Management,
https://doi.org/10.1007/978-3-031-27177-9

12. Beeri, C., Dowd, M., Fagin, R., and Statman, R. (1984). On the structure of armstrong relations for functional dependencies. *J. ACM*, 31(1):30–46.

13. Beeri, C., Fagin, R., and Howard, J. H. (1977). A complete axiomatization for functional and multivalued dependencies in database relations. In *Proceedings of the 1977 ACM SIGMOD International Conference on Management of Data, Toronto, Canada, August 3-5, 1977*, pages 47–61.

14. Beeri, C., Fagin, R., Maier, D., and Yannakakis, M. (1983). On the desirability of acyclic database schemes. *J. ACM*, 30(3):479–513.

15. Beeri, C. and Vardi, M. Y. (1984). Formal systems for tuple and equality generating dependencies. *SIAM J. Comput.*, 13(1):76–98.

16. Bertossi, L. E. (2011). *Database Repairing and Consistent Query Answering*. Synthesis Lectures on Data Management. Morgan & Claypool Publishers.

17. Bertossi, L. E., Bravo, L., Franconi, E., and Lopatenko, A. (2005). Complexity and approximation of fixing numerical attributes in databases under integrity constraints. In *DBPL*, volume 3774 of *Lecture Notes in Computer Science*, pages 262–278. Springer.

18. Bertossi, L. E., Bravo, L., Franconi, E., and Lopatenko, A. (2008). The complexity and approximation of fixing numerical attributes in databases under integrity constraints. *Inf. Syst.*, 33(4-5):407–434.

19. Bertossi, L. E. and Geerts, F. (2020). Data quality and explainable AI. *ACM J. Data Inf. Qual.*, 12(2):11:1–11:9.

20. Bertossi, L. E., Kolahi, S., and Lakshmanan, L. V. S. (2013). Data cleaning and query answering with matching dependencies and matching functions. *Theory Comput. Syst.*, 52(3):441–482.

21. Bleifuß, T., Kruse, S., and Naumann, F. (2017). Efficient denial constraint discovery with hydra. *PVLDB*, 11(3):311–323.

22. Bohannon, P., Fan, W., Geerts, F., Jia, X., and Kementsietsidis, A. (2007). Conditional functional dependencies for data cleaning. In *Proceedings of the 23rd International Conference on Data Engineering, ICDE 2007, The Marmara Hotel, Istanbul, Turkey, April 15-20, 2007*, pages 746–755.

23. Bohannon, P., Flaster, M., Fan, W., and Rastogi, R. (2005). A cost-based model and effective heuristic for repairing constraints by value modification. In *Proceedings of the ACM SIGMOD International Conference on Management of Data, Baltimore, Maryland, USA, June 14-16, 2005*, pages 143–154.

24. Bosc, P., Dubois, D., and Prade, H. (1998). Fuzzy functional dependencies and redundancy elimination. *JASIS*, 49(3):217–235.

25. Bravo, L., Fan, W., Geerts, F., and Ma, S. (2008). Increasing the expressivity of conditional functional dependencies without extra complexity. In *Proceedings of the 24th International Conference on Data Engineering, ICDE 2008, April 7-12, 2008, Cancún, México*, pages 516–525.

26. Bravo, L., Fan, W., and Ma, S. (2007). Extending dependencies with conditions. In *Proceedings of the 33rd International Conference on Very Large Data Bases, University of Vienna, Austria, September 23-27, 2007*, pages 243–254.

27. Bray, T., Paoli, J., and Sperberg-McQueen, C. M. (1997). Extensible markup language (XML). *World Wide Web J.*, 2(4):27–66.

28. Broder, A. Z., Das, S., Fontoura, M., Ghosh, B., Josifovski, V., Shanmugasundaram, J., and Vassilvitskii, S. (2011). Efficiently evaluating graph constraints in content-based publish/subscribe. In *Proceedings of the 20th International Conference on World Wide Web, WWW 2011, Hyderabad, India, March 28 - April 1, 2011*, pages 497–506.

29. Calautti, M., Greco, S., Molinaro, C., and Trubitsyna, I. (2016). Exploiting equality generating dependencies in checking chase termination. *Proc. VLDB Endow.*, 9(5):396–407.

30. Calì, A. and Pieris, A. (2011). On equality-generating dependencies in ontology querying - preliminary report. In *AMW*, volume 749 of *CEUR Workshop Proceedings*. CEUR-WS.org.

31. Caruccio, L., Deufemia, V., and Polese, G. (2016). Relaxed functional dependencies - A survey of approaches. *IEEE Trans. Knowl. Data Eng.*, 28(1):147–165.

32. Casanova, M. A., Fagin, R., and Papadimitriou, C. H. (1982). Inclusion dependencies and their interaction with functional dependencies. In *Proceedings of the ACM Symposium on Principles of Database Systems, March 29-31, 1982, Los Angeles, California, USA*, pages 171–176.

33. Chakravarthy, U. S., Grant, J., and Minker, J. (1990). Logic-based approach to semantic query optimization. *ACM Trans. Database Syst.*, 15(2):162–207.

34. Cheng, R., Chen, J., and Xie, X. (2008). Cleaning uncertain data with quality guarantees. *PVLDB*, 1(1):722–735.

35. Cheng, R. and Prabhakar, S. (2003). Managing uncertainty in sensor database. *SIGMOD Record*, 32(4):41–46.

36. Chiang, F. and Miller, R. J. (2008). Discovering data quality rules. *PVLDB*, 1(1):1166–1177.

37. Chomicki, J. and Marcinkowski, J. (2005). Minimal-change integrity maintenance using tuple deletions. *Inf. Comput.*, 197(1-2):90–121.

38. Chu, X., Ilyas, I. F., and Papotti, P. (2013a). Discovering denial constraints. *PVLDB*, 6(13):1498–1509.

39. Chu, X., Ilyas, I. F., and Papotti, P. (2013b). Holistic data cleaning: Putting violations into context. In *29th IEEE International Conference on Data Engineering, ICDE 2013, Brisbane, Australia, April 8-12, 2013*, pages 458–469.

40. Chu, X., Ilyas, I. F., Papotti, P., and Ye, Y. (2014). Ruleminer: Data quality rules discovery. In *IEEE 30th International Conference on Data Engineering, Chicago, ICDE 2014, IL, USA, March 31 - April 4, 2014*, pages 1222–1225.

41. Ciaccia, P., Golfarelli, M., and Rizzi, S. (2013). Efficient derivation of numerical dependencies. *Inf. Syst.*, 38(3):410–429.

42. Codd, E. F. (1971). Further normalization of the data base relational model. *IBM Research Report, San Jose, California*, RJ909.

43. Codd, E. F. (1974). Recent investigations in relational data base systems. In *IFIP Congress*, pages 1017–1021. North-Holland.

44. Cong, G., Fan, W., Geerts, F., Jia, X., and Ma, S. (2007). Improving data quality: Consistency and accuracy. In *Proceedings of the 33rd International Conference on Very Large Data Bases, University of Vienna, Austria, September 23-27, 2007*, pages 315–326.

45. Curé, O. (2012). Improving the data quality of drug databases using conditional dependencies and ontologies. *J. Data and Information Quality*, 4(1):3.

46. Dasu, T., Duan, R., and Srivastava, D. (2016). Data quality for temporal streams. *IEEE Data Eng. Bull.*, 39(2):78–92.

47. Dechter, R., Meiri, I., and Pearl, J. (1991). Temporal constraint networks. *Artif. Intell.*, 49(1-3):61–95.

48. Delobel, C. (1978). Normalization and hierarchical dependencies in the relational data model. *ACM Trans. Database Syst.*, 3(3):201–222.

49. Deutsch, A. (2009). FOL modeling of integrity constraints (dependencies). In *Encyclopedia of Database Systems*, pages 1155–1161. Springer US.

50. Dong, J. and Hull, R. (1982). Applying approximate order dependency to reduce indexing space. In *Proceedings of the 1982 ACM SIGMOD International Conference on Management of Data, Orlando, Florida, June 2-4, 1982.*, pages 119–127.

51. Dong, X. L. and Srivastava, D. (2013). Big data integration. *Proc. VLDB Endow.*, 6(11):1188–1189.

52. Dürsch, F., Stebner, A., Windheuser, F., Fischer, M., Friedrich, T., Strelow, N., Bleifuß, T., Harmouch, H., Jiang, L., Papenbrock, T., and Naumann, F. (2019). Inclusion dependency discovery: An experimental evaluation of thirteen algorithms. In *Proceedings of the 28th ACM International Conference on Information and Knowledge Management, CIKM 2019, Beijing, China, November 3-7, 2019*, pages 219–228.

53. Eich, M., Fender, P., and Moerkotte, G. (2016). Faster plan generation through consideration of functional dependencies and keys. *PVLDB*, 9(10):756–767.

54. Elmagarmid, A. K., Ipeirotis, P. G., and Verykios, V. S. (2007). Duplicate record detection: A survey. *IEEE Trans. Knowl. Data Eng.*, 19(1):1–16.

55. Fagin, R. (1977). Multivalued dependencies and a new normal form for relational databases. *ACM Trans. Database Syst.*, 2(3):262–278.

56. Fagin, R. (1980). Horn clauses and database dependencies (extended abstract). In *Proceedings of the 12th Annual ACM Symposium on Theory of Computing, April 28-30, 1980, Los Angeles, California, USA*, pages 123–134.

57. Fagin, R. (1981). A normal form for relational databases that is based on domians and keys. *ACM Trans. Database Syst.*, 6(3):387–415.

58. Fan, W. (2008). Dependencies revisited for improving data quality. In *Proceedings of the Twenty-Seventh ACM SIGMOD-SIGACT-SIGART Symposium on Principles of Database Systems, PODS 2008, June 9-11, 2008, Vancouver, BC, Canada*, pages 159–170.

59. Fan, W. (2019). Dependencies for graphs: Challenges and opportunities. *ACM J. Data Inf. Qual.*, 11(2):5:1–5:12.

60. Fan, W., Fan, Z., Tian, C., and Dong, X. L. (2015a). Keys for graphs. *PVLDB*, 8(12):1590–1601.

61. Fan, W., Fu, W., Jin, R., Lu, P., and Tian, C. (2022). Discovering association rules from big graphs. *Proc. VLDB Endow.*, 15(7):1479–1492.

62. Fan, W., Gao, H., Jia, X., Li, J., and Ma, S. (2011a). Dynamic constraints for record matching. *VLDB J.*, 20(4):495–520.

63. Fan, W. and Geerts, F. (2012). *Foundations of Data Quality Management*. Synthesis Lectures on Data Management. Morgan & Claypool Publishers.

64. Fan, W., Geerts, F., and Jia, X. (2009a). Conditional dependencies: A principled approach to improving data quality. In *Dataspace: The Final Frontier, 26th British National Conference on Databases, BNCOD 26, Birmingham, UK, July 7-9, 2009. Proceedings*, pages 8–20.

65. Fan, W., Geerts, F., Jia, X., and Kementsietsidis, A. (2008a). Conditional functional dependencies for capturing data inconsistencies. *ACM Trans. Database Syst.*, 33(2).

66. Fan, W., Geerts, F., Lakshmanan, L. V. S., and Xiong, M. (2009b). Discovering conditional functional dependencies. In *Proceedings of the 25th International Conference on Data Engineering, ICDE 2009, March 29 2009 - April 2 2009, Shanghai, China*, pages 1231–1234.

67. Fan, W., Geerts, F., Li, J., and Xiong, M. (2011b). Discovering conditional functional dependencies. *IEEE Trans. Knowl. Data Eng.*, 23(5):683–698.

68. Fan, W., Hu, C., Liu, X., and Lu, P. (2018). Discovering graph functional dependencies. In *Proceedings of the 2018 International Conference on Management of Data, SIGMOD Conference 2018, Houston, TX, USA, June 10-15, 2018*, pages 427–439. ACM.

69. Fan, W., Hu, C., Liu, X., and Lu, P. (2020a). Discovering graph functional dependencies. *ACM Trans. Database Syst.*, 45(3):15:1–15:42.

70. Fan, W., Jia, X., Li, J., and Ma, S. (2009c). Reasoning about record matching rules. *PVLDB*, 2(1):407–418.

71. Fan, W., Jin, R., Liu, M., Lu, P., Tian, C., and Zhou, J. (2020b). Capturing associations in graphs. *Proc. VLDB Endow.*, 13(11):1863–1876.

72. Fan, W., Li, J., Ma, S., Tang, N., and Yu, W. (2011c). Interaction between record matching and data repairing. In *Proceedings of the ACM SIGMOD International Conference on Management of Data, SIGMOD 2011, Athens, Greece, June 12-16, 2011*, pages 469–480.

73. Fan, W. and Lu, P. (2017). Dependencies for graphs. In *Proceedings of the 36th ACM SIGMOD-SIGACT-SIGAI Symposium on Principles of Database Systems, PODS 2017, Chicago, IL, USA, May 14-19, 2017*, pages 403–416.

74. Fan, W., Lu, P., and Tian, C. (2020c). Unifying logic rules and machine learning for entity enhancing. *Sci. China Inf. Sci.*, 63(7).

75. Fan, W., Ma, S., Hu, Y., Liu, J., and Wu, Y. (2008b). Propagating functional dependencies with conditions. *PVLDB*, 1(1):391–407.

76. Fan, W., Ma, S., Tang, N., and Yu, W. (2014). Interaction between record matching and data repairing. *J. Data and Information Quality*, 4(4):16:1–16:38.

77. Fan, W., Wang, X., Wu, Y., and Xu, J. (2015b). Association rules with graph patterns. *Proc. VLDB Endow.*, 8(12):1502–1513.

78. Fan, W., Wu, Y., and Xu, J. (2016). Functional dependencies for graphs. In *Proceedings of the 2016 International Conference on Management of Data, SIGMOD Conference 2016, San Francisco, CA, USA, June 26 - July 01, 2016*, pages 1843–1857.

79. Flach, P. A. and Savnik, I. (1999). Database dependency discovery: A machine learning approach. *AI Commun.*, 12(3):139–160.

80. Flesca, S., Furfaro, F., Greco, S., and Zumpano, E. (2003). Repairs and consistent answers for XML data with functional dependencies. In *Database and XML Technologies, First International XML Database Symposium, XSym 2003, Berlin, Germany, September 8, 2003, Proceedings*, pages 238–253.

81. Franklin, M. J., Halevy, A. Y., and Maier, D. (2005). From databases to dataspaces: a new abstraction for information management. *SIGMOD Record*, 34(4):27–33.

82. Gao, F., Song, S., and Wang, J. (2021). Time series data cleaning under multi-speed constraints. *Int. J. Softw. Informatics*, 11(1):29–54.

83. Ginsburg, S. and Hull, R. (1983a). Order dependency in the relational model. *Theor. Comput. Sci.*, 26:149–195.

84. Ginsburg, S. and Hull, R. (1983b). Sort sets in the relational model. In *Proceedings of the Second ACM SIGACT-SIGMOD Symposium on Principles of Database Systems, March 21-23, 1983, Colony Square Hotel, Atlanta, Georgia, USA*, pages 332–339.

85. Ginsburg, S. and Hull, R. (1986). Sort sets in the relational model. *J. ACM*, 33(3):465–488.

86. Gogacz, T. and Toruńczyk, S. (2017). Entropy bounds for conjunctive queries with functional dependencies. In *20th International Conference on Database Theory, ICDT 2017, March 21-24, 2017, Venice, Italy*, pages 15:1–15:17.

87. Golab, L., Karloff, H. J., Korn, F., Saha, A., and Srivastava, D. (2009). Sequential dependencies. *PVLDB*, 2(1):574–585.

88. Golab, L., Karloff, H. J., Korn, F., Srivastava, D., and Yu, B. (2008). On generating near-optimal tableaux for conditional functional dependencies. *PVLDB*, 1(1):376–390.

89. Grant, J. and Minker, J. (1981). Numerical dependencies. In *XP2 Workshop on Relational Database Theory, June 22-24 1981, The Pennsylvania State University, PA, USA*.

90. Gunopulos, D., Khardon, R., Mannila, H., Saluja, S., Toivonen, H., and Sharm, R. S. (2003). Discovering all most specific sentences. *ACM Trans. Database Syst.*, 28(2):140–174.

91. Halevy, A. Y., Franklin, M. J., and Maier, D. (2006). Principles of dataspace systems. In *Proceedings of the Twenty-Fifth ACM SIGACT-SIGMOD-SIGART Symposium on Principles of Database Systems, June 26-28, 2006, Chicago, Illinois, USA*, pages 1–9.

92. Hartmann, S., Köhler, H., and Link, S. (2007). Full hierarchical dependencies in fixed and undetermined universes. *Ann. Math. Artif. Intell.*, 50(1-2):195–226.

93. Huang, R., Chen, Z., Liu, Z., Song, S., and Wang, J. (2019). Tsoutlier: Explaining outliers with uniform profiles over iot data. In *2019 IEEE International Conference on Big Data (Big Data), Los Angeles, CA, USA, December 9-12, 2019*, pages 2024–2027.

94. Huang, R., Wang, J., Song, S., Lin, X., Zhu, X., and Pei, J. (2023). Efficiently cleaning structured event logs: A graph repair approach. *ACM Trans. Database Syst.*

95. Huhtala, Y., Kärkkäinen, J., Porkka, P., and Toivonen, H. (1998). Efficient discovery of functional and approximate dependencies using partitions. In *Proceedings of the Fourteenth International Conference on Data Engineering, Orlando, Florida, USA, February 23-27, 1998*, pages 392–401.

96. Huhtala, Y., Kärkkäinen, J., Porkka, P., and Toivonen, H. (1999). Tane: An efficient algorithm for discovering functional and approximate dependencies. *Comput. J.*, 42(2):100–111.

97. Ilyas, I. F. and Chu, X. (2019). *Data Cleaning*. ACM.

98. Ilyas, I. F., Markl, V., Haas, P. J., Brown, P., and Aboulnaga, A. (2004). CORDS: automatic discovery of correlations and soft functional dependencies. In *Proceedings of the ACM SIGMOD International Conference on Management of Data, Paris, France, June 13-18, 2004*, pages 647–658.

99. Intan, R. and Mukaidono, M. (2000). Fuzzy functional dependency and its application to approximate data querying. In *2000 International Database Engineering and Applications Symposium, IDEAS 2000, September 18-20, 2000, Yokohoma, Japan, Proccedings*, pages 47–54.

100. Jeffery, S. R., Alonso, G., Franklin, M. J., Hong, W., and Widom, J. (2006). Declarative support for sensor data cleaning. In *Pervasive Computing, 4th International Conference, PERVASIVE 2006, Dublin, Ireland, May 7-10, 2006, Proceedings*, pages 83–100.

101. Jensen, C. S. and Snodgrass, R. T. (1999). Temporal data management. *IEEE Trans. Knowl. Data Eng.*, 11(1):36–44.

102. Jensen, C. S., Snodgrass, R. T., and Soo, M. D. (1996). Extending existing dependency theory to temporal databases. *IEEE Trans. Knowl. Data Eng.*, 8(4):563–582.

103. Jin, R., Hong, H., Wang, H., Ruan, N., and Xiang, Y. (2010). Computing label-constraint reachability in graph databases. In *Proceedings of the ACM SIGMOD International Conference on Management of Data, SIGMOD 2010, Indianapolis, Indiana, USA, June 6-10, 2010*, pages 123–134.

104. Kang, R., Song, S., and Wang, C. (2022). Conditional regression rules. In *38th IEEE International Conference on Data Engineering, ICDE 2022, Kuala Lumpur, Malaysia, May 9-12, 2022*, pages 2481–2493. IEEE.

105. Kang, R. and Wang, H. (2021). Dynamic relation repairing for knowledge enhancement. *IEEE Transactions on Knowledge and Data Engineering*, pages 1.

106. Kenig, B., Mundra, P., Prasaad, G., Salimi, B., and Suciu, D. (2020). Mining approximate acyclic schemes from relations. In *Proceedings of the 2020 International Conference on Management of Data, SIGMOD Conference 2020, online conference [Portland, OR, USA], June 14-19, 2020*, pages 297–312.

107. Kimura, H., Huo, G., Rasin, A., Madden, S., and Zdonik, S. B. (2009). Correlation maps: A compressed access method for exploiting soft functional dependencies. *PVLDB*, 2(1):1222–1233.

108. Kivinen, J. and Mannila, H. (1992). Approximate dependency inference from relations. In *ICDT*, volume 646 of *Lecture Notes in Computer Science*, pages 86–98. Springer.

109. Kivinen, J. and Mannila, H. (1995). Approximate inference of functional dependencies from relations. *Theor. Comput. Sci.*, 149(1):129–149.

110. Kolahi, S. and Lakshmanan, L. V. S. (2009). On approximating optimum repairs for functional dependency violations. In *Database Theory - ICDT 2009, 12th International Conference, St. Petersburg, Russia, March 23-25, 2009, Proceedings*, pages 53–62.

111. Korn, F., Muthukrishnan, S., and Zhu, Y. (2003). Checks and balances: Monitoring data quality problems in network traffic databases. In *VLDB 2003, Proceedings of 29th International Conference on Very Large Data Bases, September 9-12, 2003, Berlin, Germany*, pages 536–547.

112. Koudas, N., Saha, A., Srivastava, D., and Venkatasubramanian, S. (2009). Metric functional dependencies. In *Proceedings of the 25th International Conference on Data Engineering, ICDE 2009, March 29 2009 - April 2 2009, Shanghai, China*, pages 1275–1278.

113. Kruse, S., Jentzsch, A., Papenbrock, T., Kaoudi, Z., Quiané-Ruiz, J., and Naumann, F. (2016). Rdfind: Scalable conditional inclusion dependency discovery in RDF datasets. In *Proceedings of the 2016 International Conference on Management of Data, SIGMOD Conference 2016, San Francisco, CA, USA, June 26 - July 01, 2016*, pages 953–967.

114. Kwashie, S., Liu, J., Li, J., Liu, L., Stumptner, M., and Yang, L. (2019). Certus: An effective entity resolution approach with graph differential dependencies (gdds). *PVLDB*, 12(6):653–666.

115. Kwashie, S., Liu, J., Li, J., and Ye, F. (2014). Mining differential dependencies: A subspace clustering approach. In *Databases Theory and Applications - 25th Australasian Database Conference, ADC 2014, Brisbane, QLD, Australia, July 14-16, 2014. Proceedings*, pages 50–61.

116. Kwashie, S., Liu, J., Li, J., and Ye, F. (2015). Conditional differential dependencies (cdds). In *Advances in Databases and Information Systems - 19th East European Conference, ADBIS 2015, Poitiers, France, September 8-11, 2015, Proceedings*, pages 3–17.

117. Langer, P. and Naumann, F. (2016). Efficient order dependency detection. *VLDB J.*, 25(2):223–241.

118. Levy, A. Y. and Sagiv, Y. (1995). Semantic query optimization in datalog programs. In *Proceedings of the Fourteenth ACM SIGACT-SIGMOD-SIGART Symposium on Principles of Database Systems, May 22-25, 1995, San Jose, California, USA*, pages 163–173.

119. Li, P., Szlichta, J., Böhlen, M. H., and Srivastava, D. (2020). Discovering band order dependencies. In *ICDE*, pages 1878–1881. IEEE.

120. Lian, X., Chen, L., and Song, S. (2010). Consistent query answers in inconsistent probabilistic databases. In *Proceedings of the ACM SIGMOD International Conference on Management of Data, SIGMOD 2010, Indianapolis, Indiana, USA, June 6-10, 2010*, pages 303–314.

121. Liu, J., Li, J., Liu, C., and Chen, Y. (2012). Discover dependencies from data - a review. *IEEE Trans. Knowl. Data Eng.*, 24(2):251–264.

122. Liu, Z., Zhang, Y., Huang, R., Chen, Z., Song, S., and Wang, J. (2021). EXPERIENCE: algorithms and case study for explaining repairs with uniform profiles over iot data. *ACM J. Data Inf. Qual.*, 13(3):18:1–18:17.

123. Lopatenko, A. and Bravo, L. (2007). Efficient approximation algorithms for repairing inconsistent databases. In *Proceedings of the 23rd International Conference on Data Engineering, ICDE 2007, The Marmara Hotel, Istanbul, Turkey, April 15-20, 2007*, pages 216–225.

124. Lopes, S., Petit, J., and Toumani, F. (2002). Discovering interesting inclusion dependencies: application to logical database tuning. *Inf. Syst.*, 27(1):1–19.

125. Ma, S., Duan, L., Fan, W., Hu, C., and Chen, W. (2015). Extending conditional dependencies with built-in predicates. *IEEE Trans. Knowl. Data Eng.*, 27(12):3274–3288.

126. Ma, S., Fan, W., and Bravo, L. (2014). Extending inclusion dependencies with conditions. *Theor. Comput. Sci.*, 515:64–95.

127. Ma, Z. M., Zhang, W., and Mili, F. (2002). Fuzzy data compression based on data dependencies. *Int. J. Intell. Syst.*, 17(4):409–426.

128. Mannila, H. and Räihä, K.-J. (1992). *Design of Relational Databases*. Addison-Wesley.

129. Mannila, H. and Räihä, K.-J. (1994). Algorithms for inferring functional dependencies from relations. *Data Knowl. Eng.*, 12(1):83–99.

130. Marchi, F. D. and Petit, J. (2005). Approximating a set of approximate inclusion dependencies. In *Intelligent Information Processing and Web Mining, Proceedings of the International IIS: IIPWM'05 Conference held in Gdansk, Poland, June 13-16, 2005*, pages 633–640.

131. Mei, Y., Song, S., Fang, C., Wei, Z., Fang, J., and Long, J. (2023). Discovering editing rules by deep reinforcement learning. In *IEEE International Conference on Data Engineering, ICDE*.

132. Navarro, G. (2001). A guided tour to approximate string matching. *ACM Comput. Surv.*, 33(1):31–88.

133. Ng, W. (1999a). Lexicographically ordered functional dependencies and their application to temporal relations. In *1999 International Database Engineering and Applications Symposium, IDEAS 1999, Montreal, Canada, August 2-4, 1999, Proceedings*, pages 279–287.

134. Ng, W. (1999b). Ordered functional dependencies in relational databases. *Inf. Syst.*, 24(7):535–554.

135. Ng, W. (2001). An extension of the relational data model to incorporate ordered domains. *ACM Trans. Database Syst.*, 26(3):344–383.

136. Pena, E. H. M. and de Almeida, E. C. (2018). BFASTDC: A bitwise algorithm for mining denial constraints. In *DEXA (1)*, volume 11029 of *Lecture Notes in Computer Science*, pages 53–68. Springer.

137. Prokoshyna, N., Szlichta, J., Chiang, F., Miller, R. J., and Srivastava, D. (2015). Combining quantitative and logical data cleaning. *Proc. VLDB Endow.*, 9(4):300–311.

138. Raju, K. V. S. V. N. and Majumdar, A. K. (1988). Fuzzy functional dependencies and lossless join decomposition of fuzzy relational database systems. *ACM Trans. Database Syst.*, 13(2):129–166.

139. Saha, B. and Srivastava, D. (2014). Data quality: The other face of big data. In *ICDE*, pages 1294–1297. IEEE Computer Society.

140. Salimi, B., Rodriguez, L., Howe, B., and Suciu, D. (2019). Interventional fairness: Causal database repair for algorithmic fairness. In *Proceedings of the 2019 International Conference on Management of Data, SIGMOD Conference 2019, Amsterdam, The Netherlands, June 30 - July 5, 2019*, pages 793–810.

141. Sarma, A. D., Ullman, J. D., and Widom, J. (2009). Schema design for uncertain databases. In *Proceedings of the 3rd Alberto Mendelzon International Workshop on Foundations of Data Management, Arequipa, Peru, May 12-15, 2009*.

142. Savnik, I. and Flach, P. A. (2000). Discovery of multivalued dependencies from relations. *Intell. Data Anal.*, 4(3-4):195–211.

143. Schlimmer, J. C. (1993). Efficiently inducing determinations: A complete and systematic search algorithm that uses optimal pruning. In *Machine Learning, Proceedings of the Tenth International Conference, University of Massachusetts, Amherst, MA, USA, June 27-29, 1993*, pages 284–290.

144. Song, S., Cao, Y., and Wang, J. (2016a). Cleaning timestamps with temporal constraints. *PVLDB*, 9(10):708–719.

145. Song, S. and Chen, L. (2009). Discovering matching dependencies. In *Proceedings of the 18th ACM Conference on Information and Knowledge Management, CIKM 2009, Hong Kong, China, November 2-6, 2009*, pages 1421–1424.

146. Song, S. and Chen, L. (2011). Differential dependencies: Reasoning and discovery. *ACM Trans. Database Syst.*, 36(3):16.

147. Song, S. and Chen, L. (2013). Efficient discovery of similarity constraints for matching dependencies. *Data Knowl. Eng.*, 87:146–166.

148. Song, S., Chen, L., and Cheng, H. (2012). Parameter-free determination of distance thresholds for metric distance constraints. In *IEEE 28th International Conference on Data Engineering (ICDE 2012), Washington, DC, USA (Arlington, Virginia), 1-5 April, 2012*, pages 846–857.

149. Song, S., Chen, L., and Cheng, H. (2014a). Efficient determination of distance thresholds for differential dependencies. *IEEE Trans. Knowl. Data Eng.*, 26(9):2179–2192.

150. Song, S., Chen, L., and Cheng, H. (2014b). On concise set of relative candidate keys. *PVLDB*, 7(12):1179–1190.

151. Song, S., Chen, L., and Yu, P. S. (2011). On data dependencies in dataspaces. In *Proceedings of the 27th International Conference on Data Engineering, ICDE 2011, April 11-16, 2011, Hannover, Germany*, pages 470–481.

152. Song, S., Chen, L., and Yu, P. S. (2013). Comparable dependencies over heterogeneous data. *VLDB J.*, 22(2):253–274.

153. Song, S., Cheng, H., Yu, J. X., and Chen, L. (2014c). Repairing vertex labels under neighborhood constraints. *PVLDB*, 7(11):987–998.

154. Song, S., Gao, F., Zhang, A., Wang, J., and Yu, P. (2021). Stream data cleaning under speed and acceleration constraints. *ACM Trans. Database Syst.*

155. Song, S., Liu, B., Cheng, H., Yu, J. X., and Chen, L. (2017). Graph repairing under neighborhood constraints. *VLDB J.*, 26(5):611–635.

156. Song, S., Sun, Y., Zhang, A., Chen, L., and Wang, J. (2018). Enriching data imputation under similarity rule constraints. *IEEE Transactions on Knowledge and Data Engineering*, pages 1.

157. Song, S., Zhang, A., Chen, L., and Wang, J. (2015a). Enriching data imputation with extensive similarity neighbors. *PVLDB*, 8(11):1286–1297.

158. Song, S., Zhang, A., Wang, J., and Yu, P. S. (2015b). SCREEN: stream data cleaning under speed constraints. In *Proceedings of the 2015 ACM SIGMOD International Conference on Management of Data, Melbourne, Victoria, Australia, May 31 - June 4, 2015*, pages 827–841.

159. Song, S., Zhu, H., and Wang, J. (2016b). Constraint-variance tolerant data repairing. In *Proceedings of the 2016 International Conference on Management of Data, SIGMOD Conference 2016, San Francisco, CA, USA, June 26 - July 01, 2016*, pages 877–892.

160. Szlichta, J., Godfrey, P., Golab, L., Kargar, M., and Srivastava, D. (2017). Effective and complete discovery of order dependencies via set-based axiomatization. *PVLDB*, 10(7):721–732.

161. Szlichta, J., Godfrey, P., Gryz, J., Ma, W., Qiu, W., and Zuzarte, C. (2014). Business-intelligence queries with order dependencies in DB2. In *Proceedings of the 17th International Conference on Extending Database Technology, EDBT 2014, Athens, Greece, March 24-28, 2014.*, pages 750–761.

162. Szlichta, J., Godfrey, P., Gryz, J., and Zuzarte, C. (2013). Expressiveness and complexity of order dependencies. *Proc. VLDB Endow.*, 6(14):1858–1869.

163. Thirumuruganathan, S., Berti-Équille, L., Ouzzani, M., Quiané-Ruiz, J., and Tang, N. (2017). Uguide: User-guided discovery of fd-detectable errors. In *Proceedings of the 2017 ACM International Conference on Management of Data, SIGMOD Conference 2017, Chicago, IL, USA, May 14-19, 2017*, pages 1385–1397.

164. van der Aalst, W. M. P. (2011). *Process Mining - Discovery, Conformance and Enhancement of Business Processes*. Springer.

165. Vincent, M. W., Liu, J., and Liu, C. (2004). Strong functional dependencies and their application to normal forms in XML. *ACM Trans. Database Syst.*, 29(3):445–462.

166. Vo, L. T. H., Cao, J., and Rahayu, J. W. (2011). Discovering conditional functional dependencies in XML data. In *Twenty-Second Australasian Database Conference, ADC 2011, Perth, Australia, January 2011*, pages 143–152.

167. Wang, D. Z., Dong, X. L., Sarma, A. D., Franklin, M. J., and Halevy, A. Y. (2009). Functional dependency generation and applications in pay-as-you-go data integration systems. In *12th International Workshop on the Web and Databases, WebDB 2009, Providence, Rhode Island, USA, June 28, 2009*.

168. Wang, J. (2005). Database design with equality-generating dependencies. In *DASFAA*, volume 3453 of *Lecture Notes in Computer Science*, pages 335–346. Springer.

169. Wang, J., Song, S., Lin, X., Zhu, X., and Pei, J. (2015). Cleaning structured event logs: A graph repair approach. In *31st IEEE International Conference on Data Engineering, ICDE 2015, Seoul, South Korea, April 13-17, 2015*, pages 30–41.

170. Wang, J., Song, S., Zhu, X., and Lin, X. (2013). Efficient recovery of missing events. *PVLDB*, 6(10):841–852.

171. Wang, J., Song, S., Zhu, X., Lin, X., and Sun, J. (2016a). Efficient recovery of missing events. *IEEE Trans. Knowl. Data Eng.*, 28(11):2943–2957.

172. Wang, S., Shen, J., and Hong, T. (2008). Incremental discovery of fuzzy functional dependencies. In Galindo, J., editor, *Handbook of Research on Fuzzy Information Processing in Databases*, pages 615–633. IGI Global.

173. Wang, X. and Chen, G. (2004). Discovering fuzzy functional dependencies as semantic knowledge in large databases. In *The Fourth International Conference on Electronic Business - Shaping Business Strategy in a Networked World*, pages 1136–1139.

174. Wang, Y., Song, S., and Chen, L. (2016b). A survey on accessing dataspaces. *SIGMOD Record*, 45(2):33–44.

175. Wang, Y., Song, S., Chen, L., Yu, J. X., and Cheng, H. (2017). Discovering conditional matching rules. *TKDD*, 11(4):46:1–46:38.

176. Wijsen, J. (1998). Reasoning about qualitative trends in databases. *Inf. Syst.*, 23(7):463–487.

177. Wijsen, J. (2001). Trends in databases: Reasoning and mining. *IEEE Trans. Knowl. Data Eng.*, 13(3):426–438.

178. Wijsen, J., Vandenbulcke, J., and Olivie, H. (1993). Functional dependencies generalized for temporal databases that include object-identity. In Elmasri, R., Kouramajian, V., and Thalheim, B., editors, *Entity-Relationship Approach - ER'93, 12th International Conference on the Entity-Relationship Approach, Arlington, Texas, USA, December 15-17, 1993, Proceedings*, volume 823 of *Lecture Notes in Computer Science*, pages 99–109. Springer.

179. Wolf, G., Khatri, H., Chokshi, B., Fan, J., Chen, Y., and Kambhampati, S. (2007). Query processing over incomplete autonomous databases. In *Proceedings of the 33rd International Conference on Very Large Data Bases, University of Vienna, Austria, September 23-27, 2007*, pages 651–662.

180. Wyss, C. M., Giannella, C., and Robertson, E. L. (2001). Fastfds: A heuristic-driven, depth-first algorithm for mining functional dependencies from relation instances - extended abstract. In *Data Warehousing and Knowledge Discovery, Third International Conference, DaWaK 2001, Munich, Germany, September 5-7, 2001, Proceedings*, pages 101–110.

181. Yeh, P. Z., Puri, C. A., Wagman, M., and Easo, A. K. (2011). Accelerating the discovery of data quality rules: A case study. In *Proceedings of the Twenty-Third Conference on Innovative Applications of Artificial Intelligence, August 9-11, 2011, San Francisco, California, USA*.

182. Yu, C. and Jagadish, H. V. (2006). Efficient discovery of XML data redundancies. In *Proceedings of the 32nd International Conference on Very Large Data Bases, Seoul, Korea, September 12-15, 2006*, pages 103–114.

183. Zanzi, A. and Trombetta, A. (2014). Discovering non-constant conditional functional dependencies with built-in predicates. In *Database and Expert Systems Applications - 25th International Conference, DEXA 2014, Munich, Germany, September 1-4, 2014. Proceedings, Part I*, pages 35–49.

184. Zhang, A., Song, S., Sun, Y., and Wang, J. (2019). Learning individual models for imputation. In *35th IEEE International Conference on Data Engineering, ICDE 2019, Macao, China, April 8-11, 2019*, pages 160–171. IEEE.

185. Zhang, A., Song, S., and Wang, J. (2016). Sequential data cleaning: A statistical approach. In *Proceedings of the 2016 International Conference on Management of Data, SIGMOD Conference 2016, San Francisco, CA, USA, June 26 - July 01, 2016*, pages 909–924.

186. Zhang, J., Wang, C., Wang, J., and Yu, J. X. (2014). Inferring continuous dynamic social influence and personal preference for temporal behavior prediction. *PVLDB*, 8(3):269–280.

187. Zheng, W., Zou, L., Peng, W., Yan, X., Song, S., and Zhao, D. (2016). Semantic SPARQL similarity search over RDF knowledge graphs. *PVLDB*, 9(11):840–851.

Printed in the United States
by Baker & Taylor Publisher Services